HOLT SCIENCE & TECHNOLOGY

Assessment Item Listing for Human Body Systems and Health

HOLT, RINEHART AND WINSTON

A Harcourt Classroom Education Company

Austin • New York • Orlando • Atlanta • San Francisco • Boston • Dallas • Toronto • London

Art and Photo Credits

All work, unless otherwise noted, contributed by Holt, Rinehart and Winston.

Front cover: VCG/FPG International; (owl on cover, title page) Kim Taylor/Bruce Coleman, Inc.

Printed in the United States of America

ISBN 0-03-065512-9

1 2 3 4 5 6 082 05 04 03 02 01

CONTENTS

Introduction

The *Holt Science and Technology* Test Generator and *Assessment Item Listing*

The *Holt Science and Technology* Test Generator consists of a comprehensive bank of test items and the ExamView® Pro 3.0 software, which enables you to produce your own tests based on the items in the Test Generator and items you create yourself. Both Macintosh® and Windows® versions of the Test Generator are included on the *Holt Science and Technology* One-Stop Planner with Test Generator. Directions on pp. vi–vii of this book explain how to install the program on your computer. This *Assessment Item Listing* is a printout of all the test items in the *Holt Science and Technology* Test Generator.

ExamView Software

ExamView enables you to quickly create printed and on-line tests. You can enter your own questions in a variety of formats, including true/false, multiple choice, completion, problem, short answer, and essay. The program also allows you to customize the content and appearance of the tests you create.

Test Items

The *Holt Science and Technology* Test Generator contains a file of test items for each chapter of the textbook. Each item is correlated to the chapter objectives in the textbook and by difficulty level.

Item Codes

As you browse through this *Assessment Item Listing*, you will see that all test items of the same type appear under an identifying head. Each item is coded to assist you with item selection. Following is an explanation of the codes.

MULTIPLE CHOICE

1. The amount of force exerted on a given area is called

   a. pressure. c. energy.
   b. viscosity. d. temperature.

   ANS: A DIF: I OBJ: 2-1.4

DIF defines the difficulty of the item.
I requires recall of information.
II requires analysis and interpretation of known information.
III requires application of knowledge to new situations.

OBJ lists the chapter number, section number, and objective.
(2-1.4 = Chapter 2, Section 1, Objective 4)

INSTALLATION AND STARTUP

The Test Generator is provided on the One-Stop Planner. The Test Generator includes ExamView and all of the questions for the corresponding textbook. ExamView includes three components: Test Builder, Question Bank Editor, and Test Player. The Test Builder includes options to create, edit, print, and save tests. The Question Bank Editor lets you create or edit question banks. The Test Player is a separate program that your students can use to take on-line* (computerized or LAN-based) tests. Please refer to the ExamView User's Guide on the One-Stop Planner for complete instructions.

Before you can use the Test Generator, you must install ExamView and the test banks on your hard drive. The system requirements, installation instructions, and startup procedures are provided below.

SYSTEM REQUIREMENTS

To use ExamView, your computer must meet or exceed the following hardware requirements:

Windows®

- Pentium processor
- Windows 95®, Windows 98®, Windows 2000® (or a more recent version)
- color monitor (VGA-compatible)
- CD-ROM and/or high-density floppy disk drive
- hard drive with at least 7 MB space available
- 8 MB available memory (16 MB memory recommended)
- an Internet connection (if you wish to access the Internet testing features)*

Macintosh®

- PowerPC processor, 100 MHz
- System 7.5 (or a more recent version)
- color monitor (VGA-compatible)
- CD-ROM and/or high-density floppy disk drive
- hard drive with at least 7 MB space available
- 8 MB available memory (16 MB memory recommended)
- an Internet connection with System 8.6 or a more recent version (if you wish to access the Internet testing features)*

* You can use the Test Player to host tests on your personal or school Web site or local area network (LAN) at no additional charge. The ExamView Web site's Internet test-hosting service must be purchased separately. Visit www.examview.com to learn more.

INSTALLATION

Instructions for installing ExamView from the CD-ROM:

Windows®

Step 1

Turn on your computer.

Step 2

Insert the One-Stop Planner into the CD-ROM drive.

Step 3

Click the Start button on the taskbar, and choose the Run option.

Step 4

In the Open box, type "d:\setup.exe" (substitute the letter for your drive if it is not d:) and click OK.

Step 5

Follow the prompts on the screen to complete the installation process.

Macintosh®

Step 1

Turn on your computer.

Step 2

Insert the One-Stop Planner into the CD-ROM drive. When the CD-ROM icon appears on the desktop, double-click the icon.

Step 3

Double-click the ExamView Pro Installer icon.

Step 4

Follow the prompts on the screen to complete the installation process.

Instructions for installing ExamView from the Main Menu of the One-Stop Planner (Macintosh® or Windows®):

Follow steps 1 and 2 from above.

Step 3

Double-click One-Stop.pdf. (If you do not have Adobe Acrobat® Reader installed on your computer, install it before proceeding by clicking Reader Installer.)

Step 4

To advance to the Main Menu, click anywhere on the title screen.

Step 5

Click the Test Generator button.

Step 6

Click the appropriate Install ExamView button.

Step 7

Follow the prompts on the screen to complete the installation process.

GETTING STARTED

After you complete the installation process, follow these instructions to start ExamView. See the ExamView User's Guide on the One-Stop Planner for further instructions on the program's options for creating a test and editing a question bank.

Startup Instructions

Step 1
Turn on the computer.
Step 2
Windows®: Click the Start button on the taskbar. Highlight the Programs menu, and locate the ExamView Pro Test Generator folder. Select the ExamView Pro option to start the software.
Macintosh®: Locate and open the ExamView Pro folder. Double-click the ExamView Pro icon.
Step 3
The first time you run the software, you will be prompted to enter your name, school/institution name, and city/state. You are now ready to begin using ExamView.
Step 4
Each time you start ExamView, the Startup menu appears. Choose one of the options shown.
Step 5
Use ExamView to create a test or edit questions in a question bank.

Technical Support

If you have any questions about the Test Generator, call the Holt, Rinehart and Winston technical support line at 1-800-323-9239, Monday through Friday, 7:00 A.M. to 6:00 P.M., Central Standard Time. You can contact the Technical Support Center on the Internet at http://www.hrwtechsupport.com or by e-mail at tsc@hrwtechsupport.com.

TRUE/FALSE

1. Homeostasis is the maintenance of a stable internal environment.

ANS: T DIF: I OBJ: 1-1.1

2. Epithelial tissue sends electrical signals throughout the body.

ANS: F DIF: I OBJ: 1-1.1

3. Blood is a type of connective tissue.

ANS: T DIF: I OBJ: 1-1.1

4. The lymphatic system returns leaked fluids to blood vessels.

ANS: T DIF: I OBJ: 1-1.3

MULTIPLE CHOICE

1. Which of the following is made up of cells that can contract and relax?
a. skeletal tissue
b. muscle tissue
c. connective tissue
d. nervous tissue

ANS: B DIF: I OBJ: 1-1.1

2. The organ system that provides support and protection for body parts is the
a. endocrine system.
b. circulatory system.
c. skeletal system.
d. respiratory system.

ANS: C DIF: I OBJ: 1-1.3

3. The epidermis is composed of
a. dermis.
b. epithelial tissue.
c. connective tissue.
d. true skin.

ANS: B DIF: I OBJ: 1-4.2

4. The fixed point in a lever is the
a. effort.
b. load.
c. fulcrum.
d. mechanical advantage.

ANS: C DIF: I OBJ: 1-2.5

5. Muscles cause bones to move when
   a. the muscles stretch.
   b. the muscles grow between bones.
   c. the muscles pull on bones.
   d. the muscles push bones apart.

   ANS: C DIF: I OBJ: 1-3.3

6. Ligaments are the connective tissue that attaches
   a. bones to muscles.
   b. bones to other bones.
   c. muscles to other muscles.
   d. muscles to dermis.

   ANS: B DIF: I OBJ: 1-2.4

7. The heart is able to continuously pump blood because it contains a layer of ____ muscle.
   a. smooth
   b. voluntary
   c. cardiac
   d. aerobic

   ANS: C DIF: I OBJ: 1-3.2

8. Which type of muscle always straightens a part of your body when it contracts?
   a. resistance
   b. flexor
   c. involuntary
   d. extensor

   ANS: D DIF: I OBJ: 1-3.2

9. Mature bone is composed of hard ____ tissue.
   a. connective
   b. epithelial
   c. cartilage
   d. skeletal

   ANS: A DIF: I OBJ: 1-2.3

10. Which of the following is a true statement about joints?
    a. The knee is a ball-and-socket joint.
    b. The spinal column contains no joints.
    c. The skull contains joints.
    d. They are kept together by elastic bands called tendons.

    ANS: C DIF: I OBJ: 1-2.4

11. A mole is more likely to be cancerous if it
    a. remains the same size.
    b. has two halves that do not match.
    c. forms inside a hair follicle.
    d. contains dead, keratin-filled cells.

    ANS: B DIF: I OBJ: 1-4.4

12. Epithelial tissue does NOT contain
    a. sweat glands.
    b. hair follicles.
    c. melanin.
    d. ligaments.

    ANS: D DIF: II OBJ: 1-4.2

13. The most effective form of exercise for building the size and strength of your skeletal muscles is
    a. resistance exercise.
    b. aerobic exercise.
    c. anaerobic exercise.
    d. circulatory exercise.

    ANS: A DIF: I OBJ: 1-3.4

14. The ability for the body to maintain a stable internal environment even though it is very cold outside is called
    a. homeopathy.
    b. homeostasis.
    c. homeomorphism.
    d. home control.

    ANS: B DIF: I OBJ: 1-1.1

15. Which of the following are the major tissues found in the body?
    a. epicranial, dermal, muscle, and connective tissues
    b. dermal, epicotyl, blood, and connective tissues
    c. epithelial, nervous, muscle, and connective tissues
    d. epineural, neural, dermal, and muscle tissues

    ANS: C DIF: I OBJ: 1-1.1

16. ____ tissue covers and protects underlying tissue.
    a. Connective
    b. Muscle
    c. Nervous
    d. Epithelial

    ANS: D DIF: I OBJ: 1-1.1

17. ____ tissue joins, supports, protects, insulates, nourishes, and cushions organs.
    a. Connective
    b. Muscle
    c. Nervous
    d. Epithelial

    ANS: A DIF: I OBJ: 1-1.1

18. When you look at the surface of your skin, you are looking at ____ tissue.
    a. epithelial
    b. nervous
    c. muscle
    d. connective

    ANS: A DIF: I OBJ: 1-1.1

19. ____ tissue is found in the brain, nerves, and sense organs.
    a. Connective
    b. Epithelial
    c. Nervous
    d. Muscle

    ANS: C DIF: I OBJ: 1-1.1

20. ____ tissue keeps organs from falling apart.
    a. Muscle
    b. Connective
    c. Epithelial
    d. Nervous

    ANS: B DIF: I OBJ: 1-1.1

21. Two or more tissues working together form an

a. organ.
b. organ system.
c. opus.
d. opuscule.

ANS: A DIF: I OBJ: 1-1.2

22. The stomach is an example of

a. an organ.
b. an organ system.
c. an organism.
d. a tissue.

ANS: A DIF: I OBJ: 1-1.2

23. Your body has 11 major ____ that carry out the functions of your body.

a. cells
b. tissues
c. organ systems
d. organs

ANS: C DIF: I OBJ: 1-1.2

24. Collagen is an example of

a. epithelial tissue.
b. nervous tissue.
c. muscle tissue.
d. connective tissue.

ANS: D DIF: II OBJ: 1-1.1

25. Your skin, hair, and nails protect underlying tissue and make up the ____ system.

a. epidermal
b. integumentary
c. lymphatic
d. endocrine

ANS: B DIF: I OBJ: 1-1.3

26. The ____ system is responsible for moving your bones.

a. integumentary
b. muscular
c. skeletal
d. cardiovascular

ANS: B DIF: I OBJ: 1-1.3

27. Your ____ system involves your heart pumping blood through your blood vessels.

a. respiratory
b. muscular
c. cardiovascular
d. lymphatic

ANS: C DIF: I OBJ: 1-1.3

28. It is the role of the ____ to receive and send electrical messages throughout the body.

a. nervous system
b. lymphatic system
c. endocrine system
d. integumentary system

ANS: A DIF: I OBJ: 1-1.3

29. Your ____ returns leaked fluids to blood vessels and helps you get rid of germs that can harm you.

a. urinary system
b. lymphatic system
c. endocrine system
d. digestive system

ANS: B DIF: I OBJ: 1-1.3

30. Your ____ breaks down the food you eat into nutrients that can be absorbed into your body.

a. endocrine system
b. lymphatic system
c. digestive system
d. urinary system

ANS: C DIF: I OBJ: 1-1.3

31. The ____ system involves glands regulating body functions by sending out chemical messengers.

a. nervous
b. integumentary
c. reproductive
d. endocrine

ANS: D DIF: I OBJ: 1-1.3

32. The ovaries, in females, are part of the ____ system.

a. endocrine
b. lymphatic
c. nervous
d. None of the above

ANS: A DIF: I OBJ: 1-1.3

33. The testes, in males, are part of the ____ system.

a. nervous
b. endocrine
c. lymphatic
d. None of the above

ANS: B DIF: I OBJ: 1-1.3

34. The ____ system involves your lungs absorbing oxygen and releasing carbon dioxide.

a. nervous
b. cardiovascular
c. lymphatic
d. respiratory

ANS: D DIF: I OBJ: 1-1.3

35. Your ____ system removes wastes from the blood and regulates the body's fluids.

a. endocrine
b. urinary
c. cardiovascular
d. lymphatic

ANS: B DIF: I OBJ: 1-1.3

36. Which system produces and delivers sperm?

a. lymphatic system
b. endocrine system
c. reproductive system of a male
d. reproductive system of a female

ANS: C DIF: I OBJ: 1-1.3

37. The ____ produces eggs and nourishes and shelters the unborn baby.

a. lymphatic system
b. endocrine system
c. reproductive system of a male
d. reproductive system of a female

ANS: D DIF: I OBJ: 1-1.3

38. An adult human has approximately 206 bones which are part of the ____ system.

a. integumentary
b. skeletal
c. endocrine
d. lymphatic

ANS: B DIF: I OBJ: 1-2.1

39. Your spinal cord is protected by your

a. skull.
b. ribs.
c. vertebrae.
d. pelvic girdle.

ANS: C DIF: I OBJ: 1-2.1

40. ____ marrow, sometimes found in spongy bone, produces red blood cells.

a. Red
b. Yellow
c. Purple
d. Blue

ANS: A DIF: I OBJ: 1-2.3

41. ____ marrow, found in the central cavity of long bones, stores fat.

a. Red
b. Blue
c. Purple
d. Yellow

ANS: D DIF: I OBJ: 1-2.3

42. Tiny canals within the ____ bone contain small blood vessels.

a. spongy
b. compact
c. cerebral
d. vertebral

ANS: B DIF: I OBJ: 1-2.3

43. ____ allow bones in the hand to glide over one another, giving some flexibility to the area.

a. Hinge joints
b. Ball-and-socket joints
c. Sliding joints
d. Rolling joints

ANS: C DIF: I OBJ: 1-2.4

44. ____ allows your arms to move freely in all directions.

a. Rolling joints
b. Ball-and-socket joints
c. Sliding joints
d. Hinge joints

ANS: B DIF: I OBJ: 1-2.4

45. ____ allows you to flex and extend your lower leg and upper arms.
a. Hinge joints
b. Sliding joints
c. Ball-and-socket joints
d. Rolling joints

ANS: A DIF: I OBJ: 1-2.4

Examine the illustrations below and answer the questions that follow.

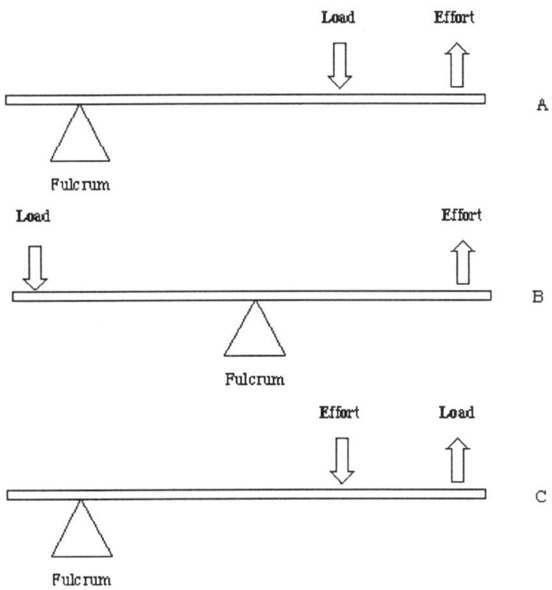

46. Which illustration is a first-class lever?
a. **A**
b. **B**
c. **C**
d. both (b) and (c)

ANS: B DIF: II OBJ: 1-2.5

47. Which illustration is a second-class lever?
a. **A**
b. **B**
c. **C**
d. both (a) and (c)

ANS: A DIF: II OBJ: 1-2.5

48. Which illustration is a third-class lever?
a. **A**
b. **B**
c. **C**
d. none of the above

ANS: C DIF: II OBJ: 1-2.5

49. Which illustration shows the lever action of your arm lifting a cup?
a. **A**
b. **B**
c. **C**
d. both (b) and (c)

ANS: C DIF: II OBJ: 1-2.5

50. Which illustration shows the lever action of your head looking at the sky?

a. **A**
b. **B**
c. **C**
d. both (b) and (c)

ANS: B DIF: II OBJ: 1-2.5

51. Which illustration shows the lever action of tiptoeing?

a. **A**
b. **B**
c. **C**
d. none of the above

ANS: A DIF: II OBJ: 1-2.5

52. Which muscle type is found in the digestive tract and in blood vessels?

a. smooth muscle
b. cardiac muscle
c. skeletal muscle
d. cranial muscle

ANS: A DIF: I OBJ: 1-3.2

53. Which muscle type is found only in your heart?

a. smooth muscle
b. skeletal muscle
c. cardiac muscle
d. cranial muscle

ANS: C DIF: I OBJ: 1-3.2

54. Which muscle type is attached to your bones for movement, and helps protect your inner organs?

a. cardiac muscle
b. skeletal muscle
c. smooth muscle
d. cranial muscle

ANS: B DIF: I OBJ: 1-3.2

55. ____ require muscles to overcome the weight of another object.

a. Aerobic exercises
b. Anaerobic exercises
c. Resistance exercises
d. Counterproductive exercises

ANS: C DIF: I OBJ: 1-3.4

56. Which action could help reduce the risk of muscle injury?

a. gradually starting an exercise program
b. warming up before strenuous exercise
c. stretching after exercise
d. all of the above

ANS: D DIF: I OBJ: 1-3.5

57. Which of the following is a function of the skin?

a. Skin bends part of your body.
b. Skin straightens out part of your body.
c. Skin helps get rid of wastes.
d. both (a) and (b)

ANS: C DIF: I OBJ: 1-4.1

58. Why do you experience a sense of touch by direct contact with your skin on objects?
a. You have moisture in your skin.
b. Your skin has sweat glands.
c. You have nerve endings in your skin.
d. Your skin has melanin.

ANS: C DIF: I OBJ: 1-4.2

59. The outer, thinner layer of the skin is called the
a. dermis.
b. follicle layer.
c. epidermis.
d. fat layer.

ANS: C DIF: I OBJ: 1-4.2

60. The deeper, thicker layer of the skin is called the
a. dermis.
b. follicle layer.
c. epidermis.
d. fat layer.

ANS: A DIF: I OBJ: 1-4.2

61. The ____ is mostly connective tissue.
a. epidermis
b. neuradermis
c. dermis
d. supadermis

ANS: C DIF: I OBJ: 1-4.2

Examine the cutaway diagram of skin below and answer the questions that follow.

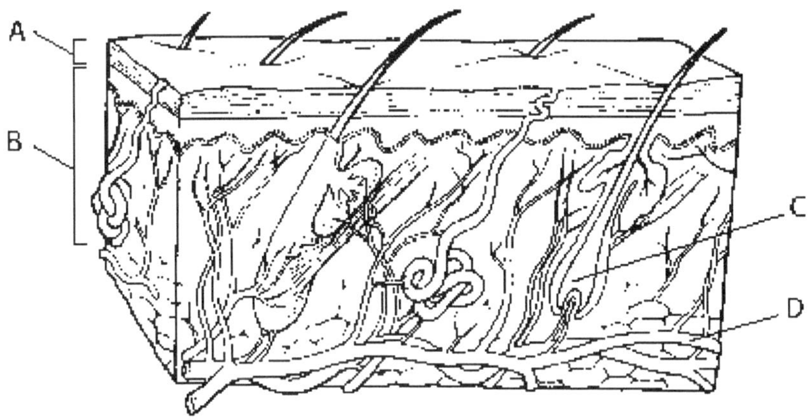

62. The epidermis is the structure labeled
a. **A**
b. **B**
c. **C**
d. **D**

ANS: A DIF: II OBJ: 1-4.2

63. The dermis is the structure labeled
a. **A**
b. **B**
c. **C**
d. **D**

ANS: B DIF: II OBJ: 1-4.2

64. The structure labeled **C** is a(n)
   a. hair follicle.
   b. blood vessel.
   c. oil gland.
   d. muscle fiber.

   ANS: A DIF: II OBJ: 1-4.2

65. The structure labeled **D** is a(n)
   a. muscle fiber.
   b. blood vessel.
   c. oil gland.
   d. hair follicle.

   ANS: B DIF: II OBJ: 1-4.2

66. A ____ is actually layers of dead, tightly packed, keratin-filled cells.
   a. nerve
   b. mole
   c. hair follicle
   d. hair

   ANS: D DIF: I OBJ: 1-4.3

67. New cells to form a ____ are produced just beneath the lunula.
   a. nail
   b. hair
   c. hair follicle
   d. mole

   ANS: A DIF: I OBJ: 1-4.3

68. Why do you NOT feel pain when you cut your hair or nails?
   a. They do not contain nerve endings.
   b. They are made of dead cells.
   c. They grow very fast.
   d. both (a) and (b)

   ANS: D DIF: I OBJ: 1-4.3

69. Which of the following contain melanin?
   a. hair
   b. skin
   c. eyes
   d. both (a) and (b)

   ANS: D DIF: I OBJ: 1-4.3

70. In children, ____ fills the center of long bones.
   a. yellow marrow
   b. fat
   c. red marrow
   d. cartilage

   ANS: C DIF: I OBJ: 1-2.3

71. In adults, ____ fills the center of long bones.
   a. cartilage
   b. fat
   c. red marrow
   d. yellow marrow

   ANS: D DIF: I OBJ: 1-2.3

Study the major bone groups illustrated on the skeleton below and answer the questions that follow.

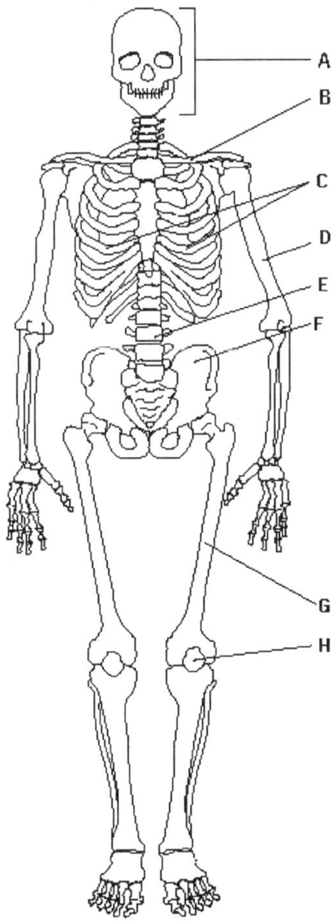

72. The structure labeled **A** in the skeleton is the
    a. skull.
    b. vertebral column.
    c. tibia.
    d. pelvic girdle.

ANS: A DIF: II OBJ: 1-2.1

73. The structure labeled **B** in the skeleton is the
    a. skull.
    b. tibia.
    c. rib.
    d. clavicle.

ANS: D DIF: II OBJ: 1-2.1

74. The structures labeled **C** in the skeleton are the
    a. vertebrae.
    b. patellas.
    c. ribs.
    d. clavicles.

ANS: C DIF: II OBJ: 1-2.1

75. The structure labeled **D** in the skeleton is the

a. humerus.
b. tibia.
c. femur.
d. clavicle.

ANS: A DIF: II OBJ: 1-2.1

76. The structure labeled **E** in the skeleton is the

a. humerus.
b. vertebral column.
c. radius.
d. pelvic girdle.

ANS: B DIF: II OBJ: 1-2.1

77. The structure labeled **F** in the skeleton is the

a. humerus.
b. vertebral column.
c. radius.
d. pelvic girdle.

ANS: D DIF: II OBJ: 1-2.1

78. The structure labeled **G** in the skeleton is the

a. femur.
b. humerus.
c. patella.
d. pelvic girdle.

ANS: A DIF: II OBJ: 1-2.1

79. The structure labeled **H** in the skeleton is the

a. femur.
b. humerus.
c. patella.
d. pelvic girdle.

ANS: C DIF: II OBJ: 1-2.1

Examine the illustrations of three different joints below, and answer the questions that follow.

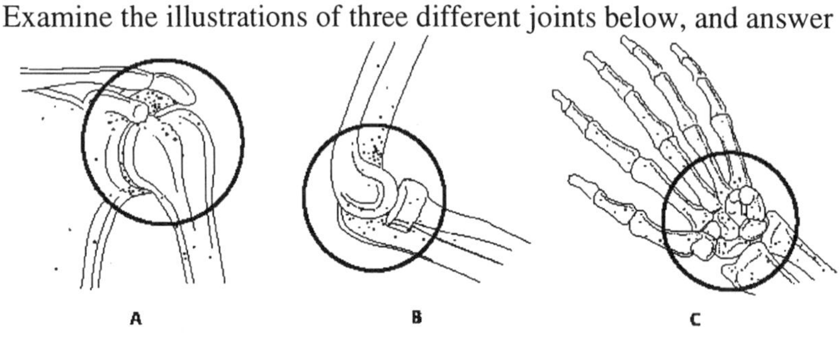

80. Which illustration shows a ball-and-socket joint?

a. **A**
b. **B**
c. **C**
d. none of the above

ANS: A DIF: II OBJ: 1-2.4

81. Joint **B** is most likely an illustration of a

a. ball-and-socket joint.
b. hinge joint.
c. gliding joint.
d. saddle joint.

ANS: B DIF: II OBJ: 1-2.4

82. Which of the joints is most likely a sliding joint?
   a. **A**
   b. **B**
   c. **C**
   d. none of the above

   ANS: C DIF: II OBJ: 1-2.4

COMPLETION

1. Electrical signals are sent throughout the body by the ____________________ tissue. (epithelial or nervous)

   ANS: nervous DIF: I OBJ: 1-1.1

2. Your ____________________ system is made up of skin, hair, and nails. (integumentary or muscular)

   ANS: integumentary DIF: I OBJ: 1-1.3

3. Bones are moved by ____________________ muscle. (smooth or skeletal)

   ANS: skeletal DIF: I OBJ: 1-3.2

4. When ____________________ muscles contract, they cause parts of the body to bend. (extensor or flexor)

   ANS: flexor DIF: I OBJ: 1-3.3

5. Most of the skeleton starts out as ____________________, which is later replaced by bone. (cartilage or ligaments)

   ANS: cartilage DIF: I OBJ: 1-2.2

6. The actions of smooth muscle and ____________________ muscle are involuntary. (skeletal or cardiac)

   ANS: cardiac DIF: I OBJ: 1-3.2

7. ____________________ bone contains marrow, which can produce red blood cells. (Spongy or Compact)

   ANS: Spongy DIF: I OBJ: 1-3.2

8. Muscles are held to bones by strong bands of tissue called ____________________. (ligaments or tendons)

   ANS: tendons DIF: I OBJ: 1-3.2

9. Most ____________________ cells are dead and filled with a protein called keratin. (dermal or epidermal)

ANS: epidermal DIF: I OBJ: 1-4.2

10. The ____________________ in skin cells determines a person's skin color. (epidermis or melanin)

ANS: melanin DIF: I OBJ: 1-4.2

11. When you look at the surface of your skin, you see ____________________ tissue. (epithelial or connective)

ANS: epithelial DIF: II OBJ: 1-4.2

12. Organs working together make up an ____________________.

ANS: organ system DIF: I OBJ: 1-1.2

13. Bones, cartilage, and the special structures that connect them make up your ____________________.

ANS: skeletal system DIF: I OBJ: 1-2.1

14. Some of your bones contain marrow that makes ____________________.

ANS: blood cells DIF: I OBJ: 1-2.3

15. Bone is composed of ________________________ and minerals that are deposited by living cells called osteoblasts.

ANS: connective tissue DIF: I OBJ: 1-2.3

16. Bone tissue that has many open spaces is called ____________________.

ANS: spongy bone DIF: I OBJ: 1-2.3

17. Bone tissue that does not have many visible open spaces is called ____________________.

ANS: compact bone DIF: I OBJ: 1-2.3

18. Bones contain a soft tissue called ____________________.

ANS: marrow DIF: I OBJ: 1-2.3

19. Most bones begin as a soft, flexible tissue called ____________________.

ANS: cartilage DIF: I OBJ: 1-2.3

20. The place where two or more bones connect is called a ____________________.

ANS: joint DIF: I OBJ: 1-2.4

21. Joints are held together with strong elastic bands of connective tissue called ____________________.

ANS: ligaments DIF: I OBJ: 1-2.4

22. Muscles that attach to bones and the connective tissue that attaches them make up the ____________________.

ANS: muscular system DIF: I OBJ: 1-3.1

23. Muscle action that is under your control is ____________________.

ANS: voluntary DIF: I OBJ: 1-3.2

24. Muscle action that is not under your control is ____________________.

ANS: involuntary DIF: I OBJ: 1-3.2

25. Strands of tough connective tissue called ____________________ connect your skeletal muscles to your bones.

ANS: tendons DIF: I OBJ: 1-3.3

26. An ____________________ is a muscle that straightens part of your body.

ANS: extensor DIF: I OBJ: 1-3.3

27. ____________________ exercises develop the size and strength of your skeletal muscles.

ANS: Resistance DIF: I OBJ: 1-3.4

28. ____________________ exercise develops the size and strength of your skeletal muscles somewhat, but mostly it strengthens the heart while increasing the endurance of your skeletal muscles.

ANS: Aerobic DIF: I OBJ: 1-3.4

29. A ____________________, commonly called a pulled muscle, is the overstretching or even tearing of a muscle.

ANS: muscle strain DIF: I OBJ: 1-3.5

30. The term ____________________ is used to describe a tumor that invades other tissue, and can be seen on the skin as a mole that has two unmatched halves.

ANS: *cancer* DIF: I OBJ: 1-4.4

31. When you get a cut, a ____________________ forms to prevent bacteria from entering the wound.

ANS: blood clot DIF: I OBJ: 1-4.4

32. A force that resists the motion of a lever, such as the downward force exerted by a weight on the bar, is called the ____________________, or the resistance.

ANS: load DIF: I OBJ: 1-2.5

SHORT ANSWER

1. Explain the relationship between cells, tissues, organs, and organ systems.

ANS:
The body is organized into cells, tissues, organs, and organ systems. Cells that have similar functions make up tissue; two or more types of tissue work together to form an organ; organs work together to form a system.

DIF: I OBJ: 1-1.2

2. Compare the four kinds of tissue found in the human body.

ANS:
Nervous tissue sends electrical signals or messages; epithelial tissue covers and protects underlying tissue; muscle tissue is made of cells that can contract and relax to produce movement; connective tissue joins, supports, protects, insulates, nourishes, cushions, and keeps organs from falling apart.

DIF: I OBJ: 1-1.1

3. Make a chart that lists the major organ systems and their functions.

ANS:
The chart should reflect the material on the 11 major organ systems found in the chapter.

DIF: II OBJ: 1-1.3

4. Describe a time when homeostasis was disrupted in your body. Which body systems do you think were affected?

   ANS:
   Answers will vary.

   DIF: II OBJ: 1-1.1

5. Describe four important functions of bones.

   ANS:
   Bones support your body, store and release minerals, and enable your muscles to move the body; some bones make blood cells.

   DIF: I OBJ: 1-2.2

6. Draw a bone, and label the inside and outside structures. Use colored pencils to color and label spongy bone, blood vessels, marrow cavity, compact bone, and cartilage.

   ANS:
   Illustrations will vary, but should reflect the structures shown in the image of the bone in the chapter.

   DIF: I OBJ: 1-2.3

7. List three hinge joints in your body.

   ANS:
   Examples include knee, knuckles, toes, elbow, and jaw.

   DIF: I OBJ: 1-2.4

8. List three types of muscle tissue, and describe their functions in the body.

   ANS:
   Smooth muscle helps move materials through the digestive tract and blood vessels; cardiac muscle causes the heart to beat; and skeletal muscle enables bones to move.

   DIF: I OBJ: 1-3.2

9. Compare aerobic exercise with resistance exercise, and give two examples of each.

   ANS:
   Resistance exercises, such as curl-ups and push-ups, build up the size and strength of skeletal muscles; they usually involve overcoming weight. Aerobic exercise, such as steady jogging, walking, or swimming, strengthens the heart while increasing endurance.

   DIF: I OBJ: 1-3.4

10. Describe the muscle action required to pick up a book. Make a sketch that illustrates the muscle action.

ANS:
Illustrations and descriptions will vary but should show that the biceps muscle shortens to bring the forearm upward, lifting the book.

DIF: II OBJ: 1-3.3

11. Why does skin color vary from person to person?

ANS:
The amount of melanin in the skin determines skin color.

DIF: I OBJ: 1-4.2

12. List six structures found in the dermis and the function of each one.

ANS:
nerves, sense; blood vessels, transport; muscle, motion; oil glands, waterproofing; sweat glands, heat regulation; fat cells, insulation.

DIF: I OBJ: 1-4.2

13. Why do you feel pain when you pull on your hair or nails but not when you cut them?

ANS:
The hair and nails do not contain nerves, but the hair follicle and nail root do.

DIF: II OBJ: 1-4.3

14. What is the difference between compact bone and spongy bone?

ANS:
Compact bone has no visible, open spaces. Spongy bone has many visible spaces and contains a tissue called marrow.

DIF: I OBJ: 1-2.3

15. Where in the body are ball-and-socket joints found?

ANS:
Ball-and-socket joints are found in the hip and shoulder.

DIF: I OBJ: 1-2.4

16. Where in the body are hinge joints found?

ANS:
Hinge joints are found in the knee and elbow.

DIF: I OBJ: 1-2.4

17. Where in the body are fixed joints found?

ANS:
Fixed joints are found in the skull.

DIF: I OBJ: 1-2.4

18. What is the difference between voluntary muscle action and involuntary muscle action? Give an example of each.

ANS:
Voluntary muscle action, such as lifting your arm, is action that you can control. Involuntary muscle action, such as the beating of your heart, is not under your control.

DIF: I OBJ: 1-3.3

19. a. What kind of muscle bends part of your body?
b. What kind of muscle straightens part of your body?

ANS:
a. A flexor bends part of my body.
b. An extensor straightens part of my body.

DIF: I OBJ: 1-3.3

20. What is the danger of using anabolic steroids?

ANS:
These powerful drugs can damage the heart, liver, and kidneys and can cause baldness. They can also cause the bones to stop growing.

DIF: I OBJ: 1-3.5

21. What are five functions of the skin?

ANS:
Skin keeps moisture in and foreign particles out, provides information about the outside world, helps to regulate body temperature, and removes some wastes.

DIF: I OBJ: 1-4.1

22. Describe the two layers of the skin.

ANS:
The epidermis is the thinner outer layer of the skin and contains epithelial tissue and mostly dead cells. The dermis, the deeper, thicker layer of skin, is made up mostly of connective tissue.

DIF: I OBJ: 1-4.2

23. How are pimples formed?

ANS:
Oil glands in the skin produce excess oil that combines with dead skin cells and bacteria to clog hair follicles and cause an infection. The result is a pimple.

DIF: I OBJ: 1-4.4

24. Summarize the functions of the four types of tissue, and draw a sketch of each type.

ANS:
Epithelial tissue—covers and protects; muscle tissue—produces movement; nervous tissue—communicates by transfer of electrical signals or messages throughout the body; connective tissue—joins, supports, and transports other tissues. (Illustrations will vary but should be similar to those in this chapter.)

DIF: I OBJ: 1-1.1

25. How does the skin help protect the body?

ANS:
Skin protects underlying tissue from dehydration, germs, and harmful radiation from the sun.

DIF: I OBJ: 1-4.2

26. What is a goose bump?

ANS:
Muscle attached to the hair follicle contracts, pushing up the epidermis as it pulls on the hair follicle.

DIF: I OBJ: 1-4.3

27. What are two ways skeletal muscle differs from cardiac muscle?

ANS:
Skeletal muscle is attached to bone by tendons and causes bones to move; cardiac muscle does not attach to bone. Skeletal muscle can be voluntary or involuntary, but cardiac muscle is involuntary.

DIF: I OBJ: 1-3.2

28. How do the functions of the skeletal system relate to the functions of the muscular system?

ANS:
Bones in your skeleton depend on muscles to move them. Bones give muscles a frame on which to move, and bones supply muscles with calcium. Bones protect muscles that carry out vital functions. The ribs protect the heart and lungs, which in turn supply the bones with oxygenated blood.

DIF: I OBJ: 1-3.3

29. Use the following terms to create a concept map: *bones, marrow, skeletal system, spongy bone, compact bone, cartilage.*

ANS:

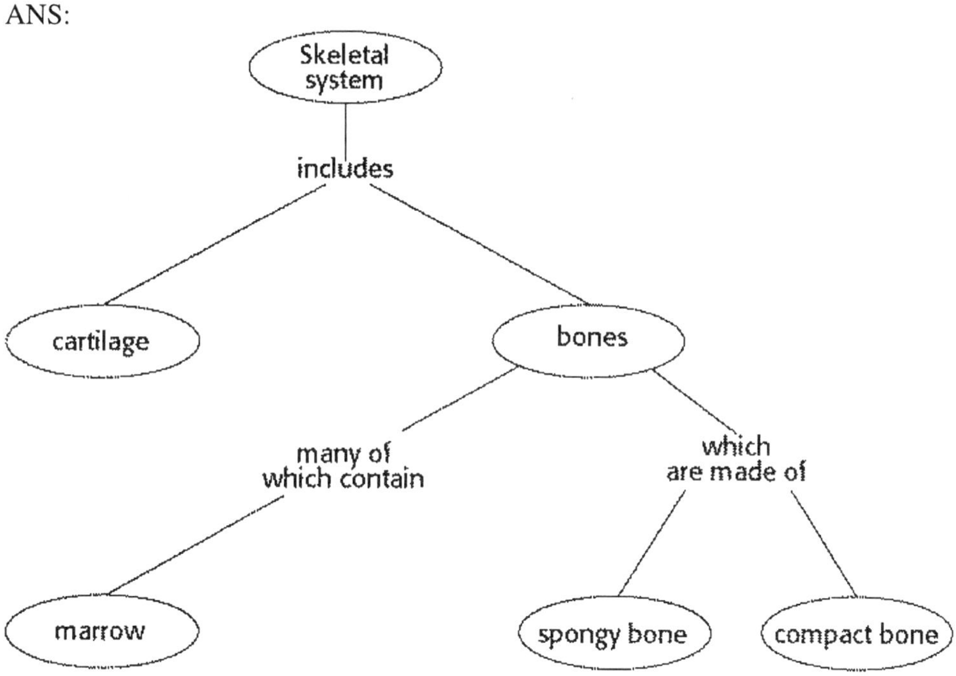

DIF: II OBJ: 1-2.3

30. Why do some muscles not work when a bone is broken?

ANS:
A muscle needs a bone for leverage. If that bone is broken in two, the muscle can only move part of the bone, or none of it depending on where the break is in relation to where the muscle's tendon attaches to the bone.

DIF: II OBJ: 1-3.3

31. Unlike human bones, some bird bones have air-filled cavities. What advantage does this give birds?

ANS:
Hollow bones are lighter, which makes flight possible.

DIF: II OBJ: 1-2.3

32. Compare the shapes of the bones of the human skull with the shapes of the bones of the human leg. Why is their shape important?

ANS:
The bones in the skull are shaped to form a strong, helmetlike protection around the brain. The leg bone has a design that makes standing upright and moving around possible.

DIF: II OBJ: 1-2.2

33. Compare the texture and sensitivity of the skin on your elbows with those of the skin on your fingertips. How can you explain the differences?

ANS:
Students should notice that the skin on their elbows is hard and rough compared with the delicate, sensitive skin on their fingertips. The skin on an elbow has to be tough and insensitive because we rest our upper body on it.

DIF: II OBJ: 1-4.2

34. Your muscles make up about 40 percent of your overall mass. What is the muscle mass of a person whose total body mass is 60 kg?

ANS:
Forty percent of 60 kg is 24 kg.

DIF: II OBJ: 1-3.1

35. The average person blinks 700 times an hour. How many times would the average person blink in a week if he or she were awake for 16 hours each day?

ANS:
700 blinks/hour × 16 hours/day × 7 days/week = 78,400 blinks/week

DIF: II OBJ: 1-3.2

36. What are the four functions of bone?

ANS:
Bone provides support and protection for the body, is the site of blood-cell formation, connects with skeletal muscle to allow for body movement, and stores minerals and fat.

DIF: I OBJ: 1-2.2

37. How does nervous tissue interact with muscle tissue?

ANS:
Nervous tissue sends electrical signals to muscle tissue to tell muscles when to contract.

DIF: I OBJ: 1-1.1

38. For a young woman, the recommended daily allowance of calcium is 800 mg per day. A pregnant woman needs 1,200 mg per day. How much more calcium per week does a pregnant woman need than a woman who is not pregnant? Show your work.

ANS:
pregnant woman: 1,200 mg/day $\times$ 7 days/week = 8,400 mg/week
woman who is not pregnant: 800 mg/day $\times$ 7 days/week = 5,600 mg/week
8,400 – 5,600 = 2,800 mg more calcium each week

DIF: II OBJ: 1-2.3

39. Use the following terms to complete the concept map below: *cartilage, tissues, organs, joints, compact bone.*

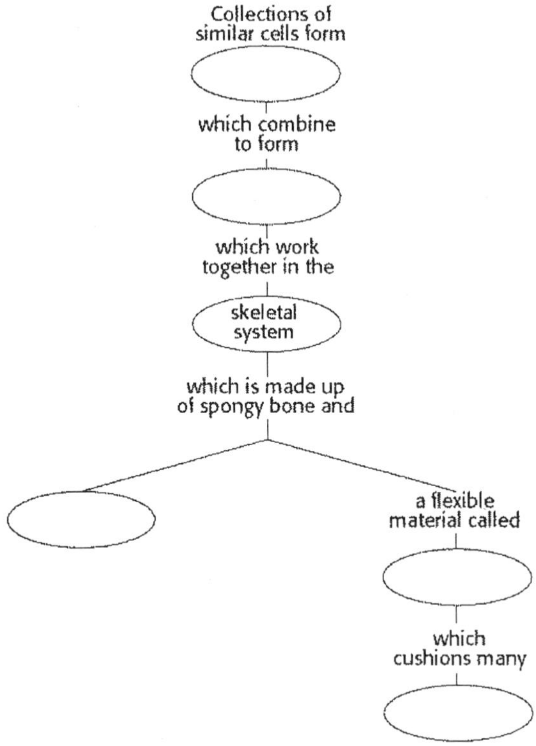

ANS:

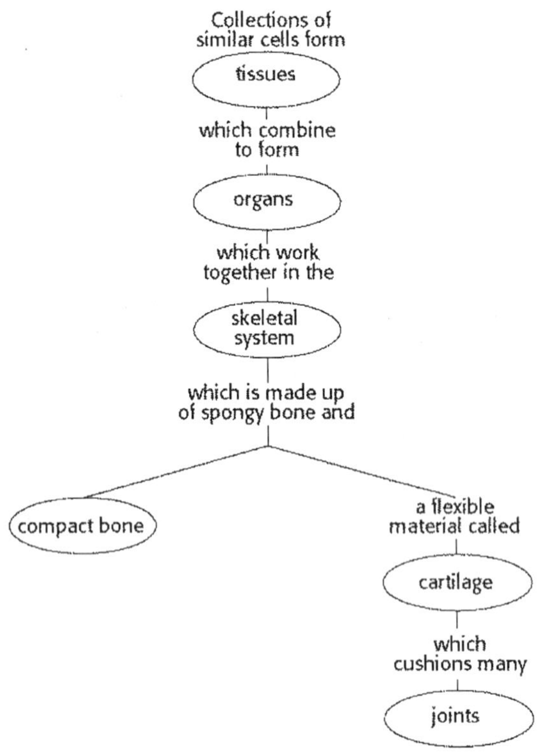

DIF: II OBJ: 1-2.3

40. Why would a doctor be particularly concerned about a person with a weak heart if the person were in a full body cast?

ANS:
A full body cast keeps a person from using their skeletal muscles. Contraction of skeletal muscle helps to move blood through the body, reducing the strain on the heart. A weak heart may not be able to move blood through the body without the help of skeletal muscles.

DIF: II OBJ: 1-3.4

The following illustration shows a cutaway view of a bone. Examine the illustration and answer the questions that follow.

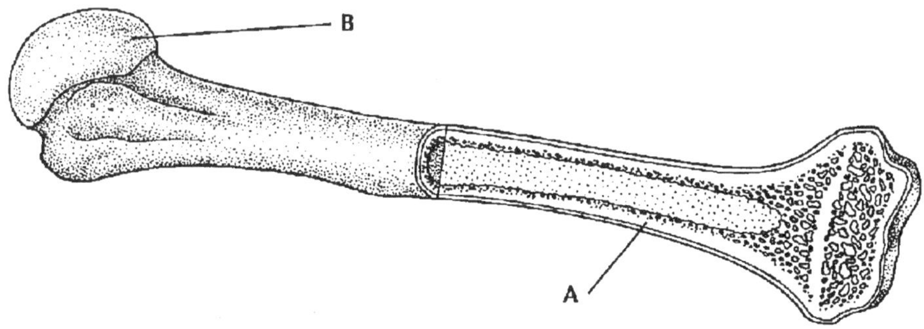

41. What is the name and function of the structure indicated by **A?**

ANS:
Compact bone provides strength and support.

DIF: II OBJ: 1-2.3

42. What is the name and function of the structure indicated by **B?**

ANS:
Cartilage covers and cushions the surface where two bones come together.

DIF: II OBJ: 1-2.3

43. How does skin help regulate body temperature?

ANS:
Sample answer: Sweat glands produce sweat, which helps cool the body.

DIF: I OBJ: 1-4.2

TRUE/FALSE

1. Type O blood is considered the universal donor because its RBCs contain all the possible antigens.

   ANS: F DIF: I OBJ: 2-1.4

2. Red blood cells are the body's first line of defense against pathogens.

   ANS: F DIF: I OBJ: 2-1.1

3. The leading cause of death in the United States is atherosclerosis.

   ANS: T DIF: I OBJ: 2-1.1

4. Arteries are blood vessels that carry blood away from the heart.

   ANS: T DIF: I OBJ: 2-1.3

5. Capillaries often are too small for red blood cells to pass through.

   ANS: F DIF: I OBJ: 2-1.2

6. Pneumonia is caused by air pollution.

   ANS: F DIF: I OBJ: 2-3.3

7. There are only two bronchi.

   ANS: T DIF: I OBJ: 2-3.1

8. The lungs are not made of any muscle.

   ANS: T DIF: I OBJ: 2-3.2

MULTIPLE CHOICE

1. Blood from the lungs enters the heart at the
   a. left ventricle.
   b. left atrium.
   c. right atrium.
   d. right ventricle.

   ANS: B DIF: I OBJ: 2-1.3

2. Blood cells are made
   a. in the heart.
   b. from plasma.
   c. from lymph.
   d. in the bones.

   ANS: D DIF: I OBJ: 2-1.1

3. Which of the following is NOT part of the lymphatic system?
   a. trachea
   b. lymph node
   c. thymus
   d. spleen

   ANS: A DIF: I OBJ: 2-2.3

4. Alveoli are surrounded by
   a. veins.
   b. muscles.
   c. capillaries.
   d. lymph nodes.

   ANS: C DIF: I OBJ: 2-3.2

5. What prevents blood from flowing backward in veins?
   a. platelets
   b. valves
   c. muscles
   d. cartilage

   ANS: B DIF: I OBJ: 2-1.2

6. Air moves into the lungs when the diaphragm muscle
   a. contracts and moves down.
   b. contracts and moves up.
   c. relaxes and moves down.
   d. relaxes and moves up.

   ANS: A DIF: I OBJ: 2-3.1

7. Oxygen-poor blood from the body returns to the ____ of the heart.
   a. left ventricle
   b. right ventricle
   c. left atrium
   d. right atrium

   ANS: D DIF: I OBJ: 2-1.3

8. Which of the following is true of people with blood type AB?
   a. Their blood contains the antibodies to A and B antigens.
   b. Half their red blood cells have A antigen, and half have B antigen.
   c. Their blood can be given to a person with blood type O.
   d. They can receive blood from people with blood type A, B, or O.

   ANS: D DIF: I OBJ: 2-1.4

9. The close arrangement of alveoli and capillaries allows
   a. damaged red blood cells to be removed from the blood.
   b. carbohydrates to move out of the blood into the lungs.
   c. oxygen to diffuse from the alveoli into the blood.
   d. carbon dioxide to diffuse from the alveoli into the blood.

   ANS: C DIF: I OBJ: 2-3.2

10. ____ is the fluid part of blood.

a. Antibody
b. Plasma
c. Hemoglobin
d. Cholesterol

ANS: B DIF: I OBJ: 2-1.1

11. Because large arteries have thick, elastic walls, they

a. are able to store blood.
b. can withstand the high pressure at which the heart pumps blood.
c. are able to collect and pump oxygen-poor blood back to the heart.
d. can support large valves.

ANS: B DIF: I OBJ: 2-1.2

12. The ____ is also called the voice box.

a. larynx
b. pharynx
c. trachea
d. bronchi

ANS: A DIF: I OBJ: 2-3.1

13. The cardiovascular system is made up of the

a. heart, lungs, and blood vessels.
b. heart, lungs, kidneys, and blood vessels.
c. heart, blood, and blood vessels.
d. heart, platelets, and blood vessels.

ANS: C DIF: I OBJ: 2-1.1

14. ____ is a connective tissue made up of two types of cells, cell parts, and plasma.

a. Collagen
b. Blood
c. Hemoglobin
d. Bone marrow

ANS: B DIF: I OBJ: 2-1.1

15. Each red blood cell contains a protein called

a. hemoglobin.
b. plasma.
c. platelets.
d. antibodies.

ANS: A DIF: I OBJ: 2-1.1

16. The most abundant cells in blood are

a. white blood cells.
b. red blood cells.
c. platelets.
d. plasma.

ANS: B DIF: I OBJ: 2-1.1

17. Which of the following is NOT lost by red blood cells just before they enter the bloodstream?

a. nucleus
b. organelles
c. hemoglobin
d. DNA

ANS: C DIF: II OBJ: 2-1.1

18. Without DNA, the red blood cells cannot
    a. contain the protein hemoglobin.
    b. carry oxygen throughout the body.
    c. be seen with a microscope.
    d. replace worn-out proteins.

    ANS: D   DIF: II   OBJ: 2-1.1

19. Some white blood cells release chemicals called ____, which help destroy bacteria, viruses, and other microscopic particles that can make you sick.
    a. platelets
    b. plasma
    c. antibodies
    d. pathogens

    ANS: C   DIF: I   OBJ: 2-1.1

20. White blood cells keep you healthy by engulfing
    a. body cells that have died.
    b. body cells that have been damaged.
    c. pathogens.
    d. All of the above

    ANS: D   DIF: I   OBJ: 2-1.1

21. When you cut or scrape your skin, you bleed because
    a. you have a surplus of red blood cells.
    b. it hurts.
    c. blood vessels have been opened.
    d. the top layer of skin has been opened.

    ANS: C   DIF: I   OBJ: 2-1.1

22. As soon as bleeding occurs, platelets
    a. begin to clump together in the damaged area.
    b. release a variety of chemicals that react with proteins in the plasma.
    c. form a plug that helps reduce blood loss.
    d. All of the above

    ANS: D   DIF: I   OBJ: 2-1.1

23. When the atria contract, blood is squeezed into
    a. arteries.
    b. veins.
    c. ventricles.
    d. capillaries.

    ANS: C   DIF: II   OBJ: 2-1.3

24. Blood from the right ventricle goes to the
    a. right atrium.
    b. left atrium.
    c. lungs.
    d. rest of the body.

    ANS: C   DIF: II   OBJ: 2-1.3

25. ____ are blood vessels that direct blood back to the heart.
    a. Arteries
    b. Capillaries
    c. Veins
    d. Platelets

    ANS: C   DIF: I   OBJ: 2-1.2

26. ____ is the pressure inside large arteries when the ventricles contract.
   a. Systemic pressure
   b. Systolic pressure
   c. Diastemic pressure
   d. Diastolic pressure

   ANS: B DIF: II OBJ: 2-1.3

27. Blood from the left ventricle goes to the
   a. right atrium.
   b. left atrium.
   c. lungs.
   d. rest of the body.

   ANS: D DIF: II OBJ: 2-1.3

28. ____ are blood vessels that direct blood away from the heart.
   a. Arteries
   b. Capillaries
   c. Veins
   d. Platelets

   ANS: A DIF: I OBJ: 2-1.2

29. ____ are the smallest blood vessels in your body.
   a. Arteries
   b. Capillaries
   c. Veins
   d. Platelets

   ANS: B DIF: I OBJ: 2-1.2

30. ____ is the pressure in large arteries when the ventricles relax.
   a. Systemic pressure
   b. Systolic pressure
   c. Diastemic pressure
   d. Diastolic pressure

   ANS: D DIF: II OBJ: 2-1.3

31. A blood pressure of 120 mm Hg means the pressure on the vessel walls is
   a. high enough to push a narrow column of mercury 120 mm high.
   b. high enough to push 120 mL of mercury to a height of 1 mm.
   c. the same as atmospheric pressure.
   d. too low for a healthy individual.

   ANS: A DIF: II OBJ: 2-1.3

32. Consistently high or low blood pressure may suggest a problem with the
   a. integumentary system.
   b. cardiovascular system.
   c. respiratory system.
   d. lymphatic system.

   ANS: B DIF: I OBJ: 2-1.3

33. When you exercise, your heart beats faster because your
   a. systolic pressure lowers too much.
   b. more red blood cells are made.
   c. muscles require much more oxygen.
   d. All of the above

   ANS: C DIF: II OBJ: 2-1.3

34. What are the four blood types?
   a. AB, O, AO, BO
   b. A, B, O, OO
   c. A, AB, B, BO
   d. A, B, AB, O

ANS: D DIF: I OBJ: 2-1.4

35. Your blood type refers to the type of ____ you have on the surface of your red blood cells.
   a. antibodies
   b. antigens
   c. platelets
   d. hemoglobin

ANS: B DIF: I OBJ: 2-1.4

36. Which antigens are found in type O blood?
   a. A antigens
   b. B antigens
   c. both A and B antigens
   d. neither A nor B antigens

ANS: D DIF: II OBJ: 2-1.4

37. A universal donor is a person with type ____ blood.
   a. A
   b. B
   c. AB
   d. O

ANS: D DIF: II OBJ: 2-1.4

38. A universal recipient is a person with type ____ blood.
   a. A
   b. B
   c. AB
   d. O

ANS: C DIF: II OBJ: 2-1.4

39. ____ in blood plasma can bind to red blood cells, causing them to clump together.
   a. Antibodies
   b. Antigens
   c. Platelets
   d. Hemoglobin

ANS: A DIF: II OBJ: 2-1.4

40. Why can a universal donor give their blood to anyone?
   a. Their red blood cells have both A and B antigens on their surface.
   b. Their red blood cells do not have any A or B antigens on their surface.
   c. Their red blood cells have only the A antigen on their surface.
   d. Their red blood cells have only the B antigen on their surface.

ANS: B DIF: II OBJ: 2-1.4

41. Your ____ helps your body fight pathogens and collects excess fluid that is not absorbed by your capillaries.
   a. integumentary system
   b. cardiovascular system
   c. respiratory system
   d. lymphatic system

ANS: D DIF: I OBJ: 2-2.1

42. The fluid and particles absorbed into lymph capillaries are called

a. blood.
b. pathogens.
c. lymph.
d. plasma.

ANS: C DIF: I OBJ: 2-2.2

43. Lymph is moved through lymphatic vessels by

a. an internal pump.
b. a lymph node.
c. the squeezing of skeletal muscles.
d. the spleen.

ANS: C DIF: II OBJ: 2-2.2

44. ____ are small bean-shaped organs where particles, such as pathogens or dead cells, are removed from the lymph.

a. Spleen
b. Lymph nodes
c. Thymus
d. Kidneys

ANS: B DIF: I OBJ: 2-2.3

45. Lymph nodes contain

a. white blood cells.
b. red blood cells.
c. thymus.
d. ventricles.

ANS: A DIF: I OBJ: 2-2.3

46. The largest lymph organ is

a. a lymph node.
b. the thymus.
c. the spleen.
d. a tonsil.

ANS: C DIF: I OBJ: 2-2.3

47. ____ is the entire process by which a body obtains and uses oxygen and gets rid of carbon dioxide and water.

a. Respiration
b. Breathing
c. Exhalation
d. Inhalation

ANS: A DIF: I OBJ: 2-3.1

48. The throat is also called the

a. pharynx.
b. larynx.
c. trachea.
d. bronchus.

ANS: A DIF: I OBJ: 2-3.1

49. The windpipe is also called the

a. pharynx.
b. larynx.
c. trachea.
d. bronchus.

ANS: C DIF: I OBJ: 2-3.1

50. The trachea splits into two tubes called
    a. the pharynx and the larynx.
    b. lungs.
    c. bronchi.
    d. alveoli.

ANS: C DIF: I OBJ: 2-3.1

51. One ____ goes to each lung and branches into thousands of tiny tubes called bronchioles.
    a. larynx
    b. bronchus
    c. pharynx
    d. alveoli

ANS: B DIF: I OBJ: 2-3.1

52. In the lungs, each bronchiole branches to form thousands of tiny sacs called
    a. tracheas.
    b. capillaries.
    c. bronchiolini.
    d. alveoli.

ANS: D DIF: I OBJ: 2-3.1

53. ____ surround each alveolus.
    a. Capillaries
    b. Lymphatic vessels
    c. Lymph nodes
    d. Bronchioles

ANS: A DIF: I OBJ: 2-3.2

54. Oxygen diffuses from the
    a. capillaries into the alveoli.
    b. alveoli into the capillaries.
    c. alveoli into the lymphatic vessels.
    d. lymphatic vessels into the alveoli.

ANS: B DIF: II OBJ: 2-3.2

55. Carbon dioxide diffuses from the
    a. capillaries into the alveoli.
    b. alveoli into the capillaries.
    c. alveoli into the lymphatic vessels.
    d. lymphatic vessels into the alveoli.

ANS: A DIF: II OBJ: 2-3.2

56. The two gases that are exchanged in the lungs are
    a. hemoglobin and platelets.
    b. oxygen and nitrogen.
    c. oxygen and carbon dioxide.
    d. oxygen and carbon monoxide.

ANS: C DIF: II OBJ: 2-3.2

57. Nicotine in cigarettes makes blood vessels become narrower. As a result, the heart
    a. does not have to work as hard to pump blood.
    b. has to work harder to pump blood.
    c. becomes weaker.
    d. becomes stronger.

ANS: B DIF: III OBJ: 2-1.1

58. Eating too much salt can lead to ____, which is abnormally high blood pressure.
   a. atherosclerosis
   b. a heart attack
   c. a stroke
   d. hypertension

   ANS: D DIF: I OBJ: 2-1.3

59. Less than 200 years ago, William Harvey discovered that blood is recycled. When he pinched a vein entering a live snake's heart, its heart would empty. When he pinched the main artery leading from the heart, the snake's heart filled with blood. Harvey thought that the blood might have a cyclic motion, as in a circle. This is why we refer to the cardiovascular system as a ____ system.
   a. circulatory
   b. respiratory
   c. dilatory
   d. pulmonatory

   ANS: A DIF: III OBJ: 2-1.1

COMPLETION

1. Oxygen is delivered to the cells of the body by ____________________. (white blood cells or red blood cells)

   ANS: red blood cells DIF: I OBJ: 2-1.1

2. Blood is carried away from the heart in ____________________. (arteries or veins)

   ANS: arteries DIF: I OBJ: 2-1.2

3. ____________________ carries nutrients to the body's cells. (Lymph or Blood)

   ANS: Blood DIF: I OBJ: 2-1.1

4. The ____________________ contains the vocal cords. (trachea or larynx)

   ANS: larynx DIF: I OBJ: 2-3.1

5. The pathway of air through the respiratory system ends at the tiny sacs called ____________________. (alveoli or bronchi)

   ANS: alveoli DIF: I OBJ: 2-3.1

6. ____________________ releases energy stored in carbohydrates, fats, and proteins. (Cellular respiration or Pulmonary circulation)

   ANS: Cellular respiration DIF: I OBJ: 2-3.1

7. The lower chambers of the heart are called ____________________. (ventricles or atria)

   ANS: ventricles DIF: I OBJ: 2-1.3

8. ____________________ circulation involves the movement of blood between the heart and the lungs. (Pulmonary or Systemic)

ANS: Pulmonary DIF: I OBJ: 2-1.3

9. A ____________________ is a small, bean-shaped organ that contains many white blood cells. (lymph node or spleen)

ANS: lymph node DIF: I OBJ: 2-2.3

10. Skeletal-muscle contraction pushes blood toward the heart through the ____________________. (arteries or veins)

ANS: veins DIF: I OBJ: 2-1.3

11. The ____________________, which releases white blood cells, is located just above the heart. (thymus or spleen)

ANS: thymus DIF: I OBJ: 2-2.3

12. The ____________________ system transports materials to and from your cells.

ANS: cardiovascular DIF: I OBJ: 2-1.1

13. ____________________ gives red blood cells their red color, and clings to the oxygen you inhale.

ANS: Hemoglobin DIF: I OBJ: 2-1.1

14. Bacteria, viruses, and other microscopic particles that can make you sick are called ____________________.

ANS: pathogens DIF: I OBJ: 2-1.1

15. Tiny particled called ____________________ are found in the blood, but they are pieces of larger cells found in bone marrow.

ANS: platelets DIF: I OBJ: 2-1.1

16. Each upper chamber of the heart is called an ____________________.

ANS: atrium DIF: I OBJ: 2-1.3

17. Each lower chamber of the heart is called a ____________________.

ANS: ventricle DIF: I OBJ: 2-1.3

18. Flaplike structures that prevent blood from flowing backward, called ____________________, are located between the atria and the ventricles and also where large arteries are attached to the heart.

ANS: valves DIF: I OBJ: 2-1.3

19. A ____________________ is a hollow tube that transports blood.

ANS: blood vessel DIF: I OBJ: 2-1.2

20. ____________________ walls are only one cell thick.

ANS: Capillary DIF: I OBJ: 2-1.2

21. The rhythmic change in blood pressure is called a ____________________.

ANS: pulse DIF: I OBJ: 2-1.3

22. ____________________ is the circulation of blood between your heart and the rest of your body other than your lungs.

ANS: Systemic circulation DIF: I OBJ: 2-1.3

23. The force exerted by blood on the inside walls of a blood vessel is called ____________________.

ANS: blood pressure DIF: I OBJ: 2-1.3

24. ____________________ occurs when fatty materials, such as cholesterol, build up on the inside of blood vessels, causing them to become narrower and less elastic.

ANS: Atherosclerosis DIF: I OBJ: 2-1.1

25. Atherosclerosis promotes ____________________, which is abnormally high blood pressure.

ANS: hypertension DIF: I OBJ: 2-1.1

26. If a blood vessel in the brain becomes clogged or ruptures, certain parts of the brain will not receive oxygen and nutrients and may die. This is called a ____________________.

ANS: stroke DIF: I OBJ: 2-1.1

27. The ____________________, located in the upper left side of your abdomen, filters blood and releases white blood cells.

ANS: spleen DIF: I OBJ: 2-2.3

28. The ____________________ system consists of the lungs, throat, and passageways that lead to the lungs.

ANS: respiratory DIF: I OBJ: 2-3.1

29. The dome-shaped muscle underneath the lungs is called the ____________________.

ANS: diaphragm DIF: I OBJ: 2-3.1

30. ____________________ is a respiratory disorder where irritants cause tissue around the bronchioles to constrict and secrete large amounts of mucus, making it difficult to breathe.

ANS: Asthma DIF: I OBJ: 2-3.3

31. ____________________ occurs when something irritates the lining of the bronchioles.

ANS: Bronchitis DIF: I OBJ: 2-3.3

32. ____________________ is a respiratory disorder that is caused by bacteria or viruses that grow inside the bronchioles and alveoli and cause them to become inflamed and filled with fluid.

ANS: Pneumonia DIF: I OBJ: 2-3.3

33. Smoking cigarettes is the leading cause of a respiratory disorder called ____________________, in which the lung tissue erodes and makes it difficult for the smoker to get oxygen.

ANS: emphysema DIF: I OBJ: 2-3.3

34. When oxygen has been absorbed by red blood cells, it is transported through the body by the ____________________ system.

ANS: cardiovascular DIF: I OBJ: 2-3.2

SHORT ANSWER

1. What is the function of the cardiovascular system?

ANS:
The cardiovascular system transports materials to and from the body's cells.

DIF: I OBJ: 2-1.1

2. What are the three kinds of blood vessels? Compare their functions.

ANS:
Arteries carry blood away from the heart to the capillaries; capillaries are small enough to allow materials in the blood to diffuse through the wall and enter the body's cells and to allow materials in the cells to diffuse into the capillaries; veins carry blood away from the capillaries back to the heart and lungs.

DIF: I OBJ: 2-1.2

3. How is the structure of capillaries related to their function?

ANS:
Capillary walls are only one cell thick. This simple structure allows nutrients, oxygen, and many other kinds of substances to diffuse easily through a capillary's wall to the body's other cells.

DIF: II OBJ: 2-1.2

4. a. Where does blood travel to and from during pulmonary circulation?
   b. Where does blood travel to and from during systemic circulation?

ANS:
a. Blood travels to and from the lungs in pulmonary circulation.
b. Blood travels to and from the body in systemic circulation.

DIF: I OBJ: 2-1.3

5. What happens to the oxygen level in blood as it moves through the lungs?

ANS:
The oxygen level increases in blood as the blood travels through the lungs.

DIF: I OBJ: 2-1.3

6. Billy has type A blood.
   a. What kind of antigens does he have on his RBCs?
   b. What blood-type antibodies can Billy make?
   c. Which blood types could be given to Billy if he needed a transfusion?

ANS:
a. A antigens
b. B antibodies
c. A and O

DIF: II OBJ: 2-1.4

7. What are the main functions of the lymphatic system?

ANS:
The lymphatic system collects excess fluid surrounding cells and returns it to the bloodstream. It also helps the body fight pathogens.

DIF: I OBJ: 2-2.1

8. Where does lymph go when it leaves the lymphatic system?

ANS:
Lymph drains into large veins in the neck.

DIF: I OBJ: 2-2.2

9. How are lymph nodes similar to the spleen?

ANS:
The spleen and lymph nodes are lymph organs that are involved in the production of white blood cells.

DIF: II OBJ: 2-2.3

10. Describe the path that air travels as it moves through the respiratory system.

ANS:
Air comes into the nose or mouth and then travels through the pharynx, larynx, trachea, and bronchi to reach the lungs.

DIF: I OBJ: 2-3.1

11. What is the difference between respiration and cellular respiration?

ANS:
Respiration involves inhaling and exhaling air, as well as cellular respiration. Cellular respiration involves the chemical reactions that release energy inside the cell.

DIF: I OBJ: 2-3.1

12. How is the function of the respiratory system related to that of the cardiovascular system?

ANS:
The respiratory system brings in oxygen and expels carbon dioxide, and the cardiovascular system transports those materials to and from the lungs.

DIF: II OBJ: 2-3.2

13. What is one function of the tonsils?

ANS:
The tonsils contain white blood cells, which help fight pathogens.

DIF: I OBJ: 2-2.3

14. How is the lymphatic system a circulatory system?

ANS:
The lymphatic system collects fluid that is not reabsorbed by the capillaries and returns it to the bloodstream.

DIF: I OBJ: 2-2.2

15. What is the difference between pulmonary circulation and systemic circulation in the cardiovascular system?

ANS:
Pulmonary circulation carries blood through the lungs and back to the heart. Systemic circulation carries blood from the heart to the rest of the body.

DIF: I OBJ: 2-1.3

16. Walton has a blood pressure of 110/65. What do the two numbers mean?

ANS:
The first number is the systolic pressure, or the pressure inside the artery when the ventricles contract. The second number, the diastolic pressure, is the pressure in the arteries when the ventricles relax.

DIF: I OBJ: 2-1.3

17. What body process produces the carbon dioxide you exhale?

ANS:
Carbon dioxide is a product of cellular respiration which occurs in the body's cells.

DIF: I OBJ: 2-3.1

18. Use the following terms to create a concept map: *blood, oxygen, alveoli, capillaries, carbon dioxide.*

ANS:

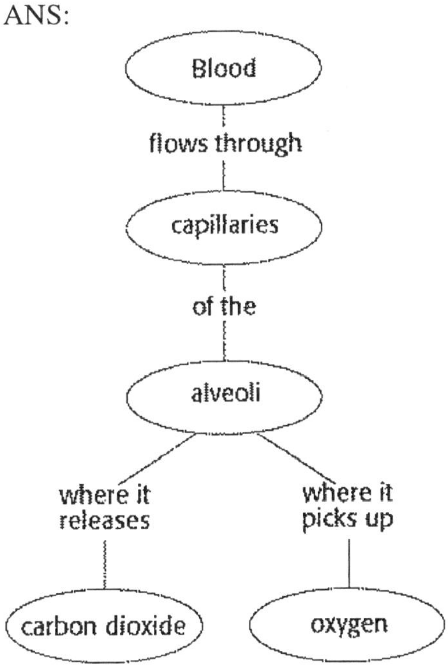

DIF: II OBJ: 2-3.2

19. Why do you think there are hairs in your nose?

ANS:
The hairs catch dust and other foreign particles. This helps keep your lungs as clean as possible.

DIF: II OBJ: 2-3.1

20. When a person is not feeling well, sometimes a doctor will examine samples of the person's blood to see how many white blood cells are present. Why would this information be useful?

ANS:
Doctors will need to know if the immune system is producing large numbers of white blood cells; this would indicate an infection.

DIF: II OBJ: 2-2.1

21. How is the function of the lymphatic system related to the function of the cardiovascular system?

ANS:
The lymphatic system returns leaked fluids back to your blood.

DIF: II OBJ: 2-2.1

22. After a person donates blood, the blood is stored in one-pint bags until it is needed for a transfusion. A healthy person normally has 5 million RBCs in each cubic millimeter (1 mm^3) of blood.
    a. How many RBCs are there in 1 mL of blood? One milliliter is equal to 1 cm^3 and to 1,000 mm^3.
    b. How many RBCs are there in 1 pt? One pint is equal to 473 mL.

ANS:
a. There are 5 billion (5,000,000,000) RBCs in 1 mL of blood.
b. There are 2.365 trillion (2,365,000,000,000) RBCs in 1 pt of blood.

DIF: II OBJ: 2-1.1

The diagram below shows how the human heart would look in cross section. Examine the diagram, and then answer the questions that follow.

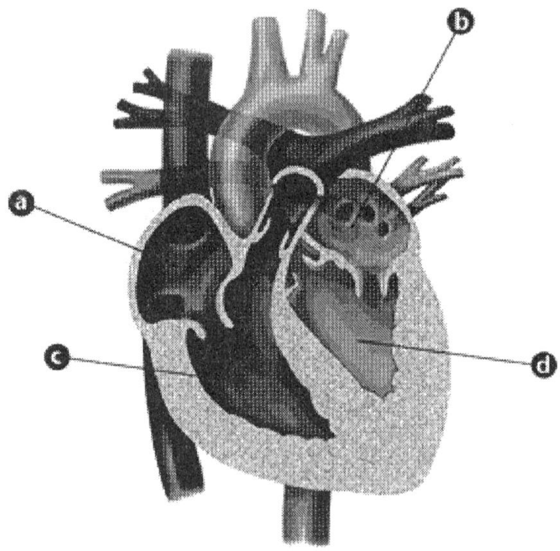

23. a. Which letter identifies the chamber that receives blood from systemic circulation?
    b. What is this chamber's name?

ANS:
a. The letter **a** identifies this chamber.
b. This chamber is called the right atrium.

DIF: II OBJ: 2-1.3

24. a. Which letter identifies the chamber that receives blood from the lungs?
    b. What is this chamber's name?

ANS:
a. The letter **b** identifies this chamber.
b. This chamber is called the left atrium.

DIF: II OBJ: 2-1.3

25. a. Which letter identifies the chamber that pumps blood to the lungs?
    b. What is this chamber's name?

ANS:
a. The letter **c** identifies this chamber.
b. This chamber is called the right ventricle.

DIF: II OBJ: 2-1.3

26. a. What characteristics of red blood cells allow them to carry oxygen?

    b. What characteristic of red blood cells allows them to flow through narrow capillaries?

ANS:

a. Red blood cells can carry oxygen because they contain a protein called hemoglobin, which binds oxygen. Additionally, their round shape gives them a large surface area for absorbing and releasing oxygen.
b. Red blood cells can flow through narrow capillaries because they do not have nuclei or other organelles. This increases their flexibility.

DIF: I OBJ: 2-1.1

27. Describe the two main functions of the lymphatic system.

ANS:
The lymphatic system collects fluid around body cells and returns this fluid to the blood through lymphatic vessels. It also fights pathogens in the body using the thymus, the spleen, the tonsils, and white blood cells.

DIF: I OBJ: 2-2.1

28. Use the following terms to complete the concept map below: *antigens, platelets, blood vessels, red blood cells, plasma, pathogens.*

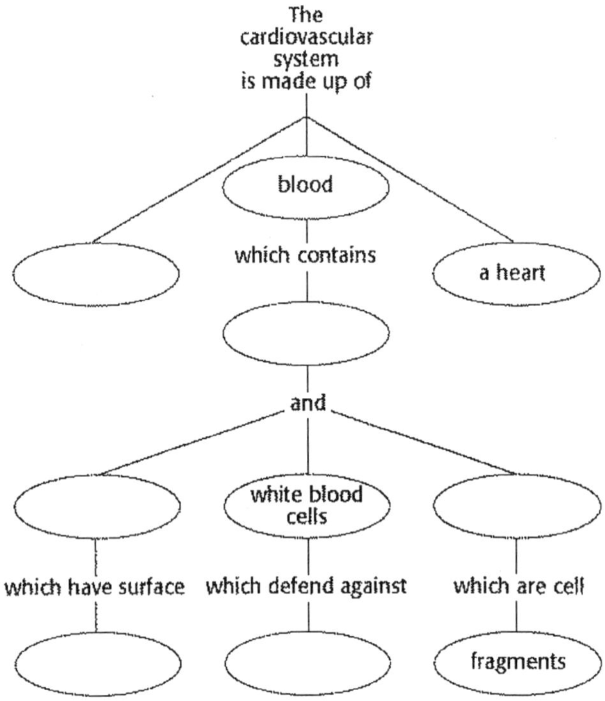

ANS:

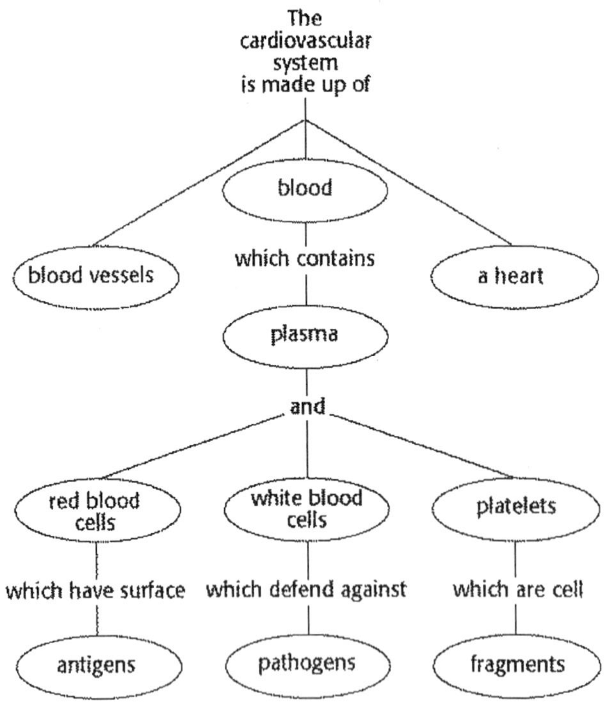

DIF: II OBJ: 2-1.1

29. Sometimes it is necessary to remove a person's spleen. How might this affect the person?

ANS:
The spleen filters blood and produces white blood cells. If a person had his or her spleen removed, he or she would not be able to recycle blood through the spleen and might have trouble fighting off infection.

DIF: II OBJ: 2-2.3

The following illustration shows the parts of the respiratory system. Examine the illustration and answer the questions that follow.

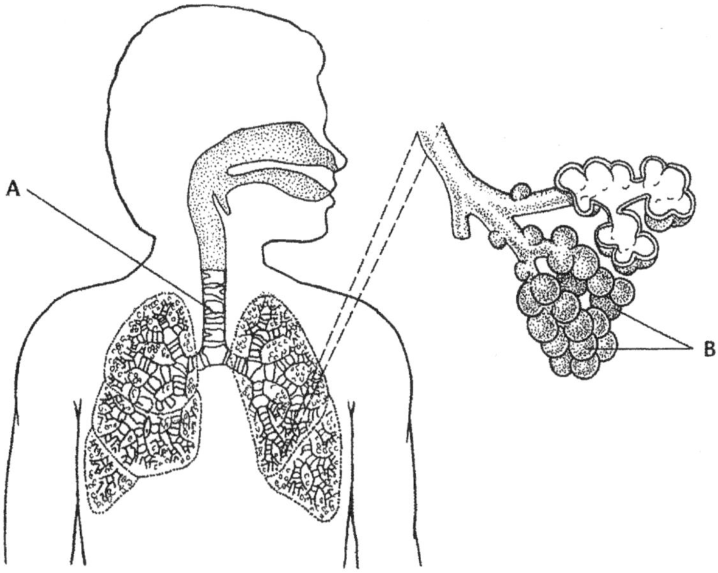

30. What is the name and the function of the body part indicated by A?

ANS:
The trachea, or windpipe, provides a passageway for air traveling from the larynx to the lungs.

DIF: II OBJ: 2-3.1

31. What is the name and the function of the body part indicated by B?

ANS:
Alveoli are tiny sacs surrounded by capillaries. They are responsible for gas exchange between the respiratory system and the blood.

DIF: II OBJ: 2-3.1

32. One of the symptoms of catching a cold or the flu is that your nose runs or becomes clogged with mucus. Explain how producing more mucus could be beneficial while you are sick, even though it makes breathing more difficult.

ANS:
Mucus in the nose traps bacteria and dust particles, preventing them from moving farther into the body. When you blow your nose, this potentially harmful debris is removed from your body.

DIF: III OBJ: 2-3.1

33. A low red blood cell count is called *anemia*. Why might people who have anemia often feel tired?

ANS:
With fewer red blood cells, blood cannot carry as much oxygen to the cells. Without enough oxygen, cells cannot break down food molecules to obtain energy efficiently, and the person feels tired.

DIF: III OBJ: 2-1.1

34. What would happen if the heart did NOT pump faster while doing exercise?

ANS:
Active muscles would not get enough oxygen, so you would get tired very quickly.

DIF: II OBJ: 2-1.3

35. As altitude increases, the gases that make up the atmosphere—including oxygen—become more scarce. When a runner who trained at sea level competes at a location 500 m above sea level, how will his performance at this altitude compare with his performance during training?

ANS:
The runner will likely experience shortness of breath and tire more quickly at the higher altitude, so his performance will be poorer.

DIF: III OBJ: 2-3.2

36. The frequency of blood clots and heart attacks is much lower among the Inuit, the nomadic hunters of the North American Arctic, than it is among other North Americans and Europeans. This difference is due to fish oils in the Inuit diet that cause their platelets to be more slippery. How would you expect the clotting ability of the Inuit's blood to be affected by the slippery platelets?

ANS:
Sample answer: I would expect the Inuit's blood to have a decreased ability to clot. The slippery platelets would be less likely to clump together in the damaged area of a blood vessel. Once a clot begins to form, slippery platelets might pass through the fibrous net, slowing the clotting process.

DIF: III OBJ: 2-1.1

TRUE/FALSE

1. Digestion begins when food reaches the stomach.

   ANS: F DIF: I OBJ: 3-1.1

2. Breaking, crushing, and mashing food are examples of chemical digestion.

   ANS: F DIF: I OBJ: 3-1.2

3. Saliva contains enzymes, which begin the chemical digestion of food.

   ANS: T DIF: I OBJ: 3-1.2

4. The esophagus connects the mouth with the small intestine.

   ANS: F DIF: I OBJ: 3-1.1

5. Undigested materials are changed from a liquid to a solid mixture in the large intestine.

   ANS: T DIF: I OBJ: 3-1.1

6. The main function of the urinary system is to remove wastes from the blood.

   ANS: T DIF: I OBJ: 3-2.1

7. The kidneys are made up of small filtering factories called ureas.

   ANS: F DIF: I OBJ: 3-2.2

8. Nephrons are responsible for forming urine.

   ANS: T DIF: I OBJ: 3-2.2

9. Sweating and urinating are two ways that the body excretes excess water.

   ANS: T DIF: I OBJ: 3-2.1

MULTIPLE CHOICE

1. The hormone that signals the kidneys to make less urine is
   a. urea. c. cellulose.
   b. ADH. d. ATP.

   ANS: B DIF: I OBJ: 3-2.1

2. Which of the following aids digestion by producing bile?
   a. stomach
   b. pancreas
   c. gallbladder
   d. liver

   ANS: D DIF: I OBJ: 3-1.1

3. The part of the kidney that filters the blood is the
   a. artery.
   b. ureter.
   c. nephron.
   d. urethra.

   ANS: C DIF: I OBJ: 3-2.2

4. The fingerlike projections lining the small intestine are called
   a. emulsifiers.
   b. fats.
   c. amino acids.
   d. villi.

   ANS: D DIF: I OBJ: 3-1.1

5. Which of the following is NOT part of the digestive tract?
   a. mouth
   b. pancreas
   c. stomach
   d. rectum

   ANS: B DIF: I OBJ: 3-1.1

6. The soupy mixture of food, enzymes, and acids in the stomach is called
   a. chyme.
   b. villi.
   c. urea.
   d. vitamins.

   ANS: A DIF: I OBJ: 3-1.1

7. Which of the following is NOT a true statement about the kidneys?
   a. Blood leaves the kidneys through the ureter.
   b. The kidneys filter about 2,000 L of blood every day.
   c. The kidneys contain microscopic filters that clean the blood.
   d. A concentrated mixture of waste materials are collected in the kidneys for excretion.

   ANS: A DIF: I OBJ: 3-2.2

8. The burning pain of indigestion is
   a. the result of gastric ulcers.
   b. caused when chyme is pushed back up in the esophagus.
   c. an initial sign of colon cancer.
   d. usually treated with antibiotics.

   ANS: B DIF: I OBJ: 3-1.3

9. Urine is stored in the ____ until it can be eliminated.
   a. urinary bladder
   b. urethra
   c. nephron
   d. ureter

   ANS: A DIF: I OBJ: 3-2.1

10. Extra nutrients absorbed by the body are stored in the
   a. colon.
   b. large intestine.
   c. liver.
   d. gall bladder.

   ANS: C DIF: I OBJ: 3-1.2

11. The lining of the small intestine is covered with villi that
   a. break fat into small droplets.
   b. contract to move food along.
   c. neutralize the acid content of chyme.
   d. increase the surface area for nutrient absorption.

   ANS: D DIF: I OBJ: 3-1.1

12. Which of the following is NOT a true statement about the pancreas?
   a. It releases bicarbonate compounds.
   b. It processes and stores bile and cholesterol.
   c. It makes endocrine hormones that regulate blood sugar.
   d. It makes enzymes that digest the major nutrients in food.

   ANS: B DIF: I OBJ: 3-1.1

13. The breaking, crushing, and mashing of food is called
   a. chemical digestion.
   b. mechanical digestion.
   c. peristaltic digestion.
   d. enzymatic digestion.

   ANS: B DIF: I OBJ: 3-1.2

14. Large molecules are broken down into nutrients in
   a. chemical digestion.
   b. peristaltic digestion.
   c. mechanical digestion.
   d. rotational digestion.

   ANS: A DIF: I OBJ: 3-1.2

15. Some nutrients are broken down into smaller particles that the body can use by
   a. amino acids.
   b. proteins.
   c. enzymes.
   d. ribosomes.

   ANS: C DIF: I OBJ: 3-1.2

16. Where does digestion begin?
   a. in the mouth
   b. in the stomach
   c. in the small intestine
   d. in the large intestine

   ANS: A DIF: I OBJ: 3-1.1

17. When you take your first bite out of an apple, which kind of teeth would you most likely be using?
   a. molars
   b. premolars
   c. enamel
   d. incisors

   ANS: D DIF: II OBJ: 3-1.1

18. Using your teeth to crush food is an example of

a. peristaltic digestion.	c. chemical digestion.
b. enzymatic digestion.	d. mechanical digestion.

ANS: D DIF: II OBJ: 3-1.2

19. Enzymes in saliva begin the process of

a. rotational digestion.	c. chemical digestion.
b. peristaltic digestion.	d. mechanical digestion.

ANS: C DIF: II OBJ: 3-1.2

20. Once the food has been reduced to a soft mush, the tongue pushes it into the throat, which leads to a long, straight tube called the

a. esophagus.	c. spleen.
b. pancreas.	d. stomach.

ANS: A DIF: I OBJ: 3-1.1

21. The muscular, baglike organ attached to the lower end of the esophagus that continues the physical digestion of a meal is called the

a. large intestine.	c. spleen.
b. small intestine.	d. stomach.

ANS: D DIF: I OBJ: 3-1.1

22. What kind of digestion does your stomach perform on your food?

a. peristaltic digestion	c. mechanical digestion
b. chemical digestion	d. Both (b) and (c)

ANS: D DIF: II OBJ: 3-1.2

23. If the food you eat contains bacteria, your stomach will most likely
    a. destroy it during chemical digestion.
    b. destroy it during mechanical digestion.
    c. pass it on to the intestines.
    d. revolt, causing you to vomit.

ANS: A DIF: II OBJ: 3-1.2

24. Chyme is slowly released into the ____ through a small ring of muscle that works like a valve.

a. large intestine	c. spleen
b. small intestine	d. stomach

ANS: B DIF: I OBJ: 3-1.1

25. The advantage of releasing chyme slowly from the stomach is that it allows more time for the chyme to mix with fluids from the

a. small and large intestines.	c. liver and pancreas.
b. gall bladder and liver.	d. spleen and liver.

ANS: C DIF: I OBJ: 3-1.1

26. Nutrients are absorbed into the bloodstream by means of
   a. nutrient-absorbing cells.
   b. fingerlike extensions in the small intestine.
   c. villi extending into the chyme.
   d. All of the above

   ANS: D DIF: I OBJ: 3-1.1

27. Which of the following is linked to the small intestine, but does NOT move food through it?
   a. liver
   b. gallbladder
   c. pancreas
   d. All of the above

   ANS: D DIF: I OBJ: 3-1.1

28. ____ contain(s) digestive enzymes and bicarbonate that neutralize the acid in chyme.
   a. Pancreatic juice
   b. Bile
   c. Stomach enzymes
   d. All of the above

   ANS: A DIF: I OBJ: 3-1.1

29. Which of the following organs makes hormones that regulate blood sugar?
   a. liver
   b. gallbladder
   c. pancreas
   d. small intestine

   ANS: C DIF: I OBJ: 3-1.1

30. Which of the following is NOT a task performed by the liver?
   a. storing nutrients
   b. absorbing nutrients into the bloodstream
   c. breaking down toxic substances in blood
   d. making cholesterol for cell membranes

   ANS: B DIF: I OBJ: 3-1.1

31. Bile is used to
   a. make cholesterol.
   b. store nutrients.
   c. break down fat.
   d. make digestive enzymes.

   ANS: C DIF: I OBJ: 3-1.2

32. When bile breaks up large fat droplets into very small droplets, it uses a ____ process.
   a. peristaltic
   b. chemical
   c. physical
   d. rotational

   ANS: C DIF: II OBJ: 3-1.2

33. Which of the following organs produces enzymes that are used to break down alcohol?
   a. small intestine
   b. pancreas
   c. gallbladder
   d. liver

   ANS: D DIF: I OBJ: 3-1.1

34. The organ of the digestive system that stores, compacts, and then eliminates indigestible material from the body is the

a. small intestine.	c. gallbladder.
b. large intestine.	d. pancreas.

ANS: B DIF: I OBJ: 3-1.1

35. Which statement best describes the function of the large intestine?
    a. It reabsorbs most of the water in the soupy mixture that enters it.
    b. It changes the soupy mixture into a solid mass.
    c. It passes cellulose to the rectum instead of the bloodstream.
    d. All of the above

ANS: D DIF: I OBJ: 3-1.1

36. Which eating behavior could cause heartburn?

a. eating too much	c. eating very acidic foods
b. eating right before going to bed	d. All of the above

ANS: D DIF: I OBJ: 3-1.3

37. Which of the following could cause constipation?

a. eating plenty of fiber	c. dried contents in the large intestine
b. drinking too much water	d. All of the above

ANS: C DIF: I OBJ: 3-1.3

38. Diarrhea occurs when bowel movements become

a. difficult.	c. frequent and watery.
b. less frequent.	d. nonexistent.

ANS: C DIF: I OBJ: 3-1.3

39. When too little water is removed from digested food in the large intestine, ____ may result.

a. heartburn	c. colon cancer
b. diarrhea	d. constipation

ANS: B DIF: I OBJ: 3-1.3

40. Which statement best describes how your body excretes waste?
    a. Your skin releases waste products and water when you sweat.
    b. Your lungs expel carbon dioxide and water when you exhale.
    c. Your urinary system removes waste products from your blood.
    d. All of the above

ANS: D DIF: I OBJ: 3-2.1

41. Which system is NOT involved in excretion?

a. the integumentary system	c. the digestive system
b. the respiratory system	d. the urinary system

ANS: C DIF: II OBJ: 3-2.1

42. The ____ are a pair of bean-shaped organs that filter about 2,000 L of blood each day.
   a. kidneys
   b. urinary bladders
   c. ureters
   d. urethras

   ANS: A   DIF: I   OBJ: 3-2.1

43. When cells use protein for energy, a by-product is a harmful substance called ____, which contains nitrogen.
   a. ureter
   b. urethra
   c. urea
   d. nephrons

   ANS: C   DIF: I   OBJ: 3-2.1

44. A(n) ____ brings blood into each kidney.
   a. vein
   b. artery
   c. lymphatic vessel
   d. tube called the ureter

   ANS: B   DIF: I   OBJ: 3-2.2

45. Within each kidney, tiny branches of blood vessels pass through part of each
   a. artery.
   b. vein.
   c. nephron.
   d. ureter.

   ANS: C   DIF: I   OBJ: 3-2.2

46. Which of the following small substances is NOT absorbed by nephrons?
   a. water
   b. urea
   c. amino acids
   d. feces

   ANS: D   DIF: I   OBJ: 3-2.2

47. Which of the following may be *reabsorbed* by blood vessels that wrap around nephrons?
   a. water
   b. nutrients
   c. concentrated waste
   d. Both (a) and (b)

   ANS: D   DIF: I   OBJ: 3-2.2

48. What is the difference between blood that enters and blood that leaves the kidneys?
   a. It is cleaned of waste material.
   b. It has a little less water.
   c. It leaves through the artery.
   d. Both (a) and (b)

   ANS: D   DIF: I   OBJ: 3-2.2

49. Urine leaves each kidney through a slender tube called the
   a. urea.
   b. ureter.
   c. urinary bladder.
   d. urethra.

   ANS: B   DIF: I   OBJ: 3-2.1

50. Urine leaves the body through a tube called the

a. urea.
b. ureter.
c. urinary bladder.
d. urethra.

ANS: D DIF: I OBJ: 3-2.1

51. Beverages containing caffeine cause the amount of water in the blood to

a. increase.
b. decrease.
c. be retained.
d. Both (a) and (c)

ANS: B DIF: I OBJ: 3-2.1

52. When you get thirsty, a hormone called ____ hormone signals the kidneys to return water to the bloodstream.

a. antidiuretic
b. diuretic
c. nephritic
d. monamine

ANS: A DIF: I OBJ: 3-2.2

53. The large intestine is NOT classified as a major excretory organ because

a. feces must pass through a membrane to be removed from the body.
b. feces does not pass through a membrane to be removed from the body.
c. it is connected directly to the stomach instead of the kidneys.
d. Both (a) and (c)

ANS: B DIF: II OBJ: 3-2.1

54. All cells produce nitrogen-containing waste products as a result of protein metabolism that enter the bloodstream. Which system of the body removes nitrogen waste products from the blood?

a. digestive system
b. urinary system
c. respiratory system
d. integumentary system

ANS: B DIF: III OBJ: 3-2.1

55. Chefs sometimes tenderize meats by either pounding the meat to a uniform thickness or rolling a device containing spikes or blades across the meat, partially cutting the meat fibers. This method of tenderization is similar to ____ in the body.

a. mechanical digestion
b. chemical digestion
c. peristaltic digestion
d. enzymatic digestion

ANS: A DIF: II OBJ: 3-1.2

56. Have you ever had chicken in a pineapple marinade? Cooks have long used the natural enzyme *bromelain* found in pineapple which contains acids that can break apart large proteins. This process of tenderizing meat is similar to ____ in the body.

a. mechanical digestion
b. chemical digestion
c. rotational digestion
d. peristaltic digestion

ANS: B DIF: II OBJ: 3-1.2

57. Overuse of salt could cause a severe affliction of the kidneys called *nephritis*. Infer what part of the kidneys you think this primarily affects.
  a. blood vessels
  b. arteries
  c. veins
  d. nephrons

  ANS: D DIF: III OBJ: 3-2.3

58. Starches (chains of sugar molecules) are broken down into sugars by enzymes called *amylases* that are in saliva. Fruits are mostly made of simple sugars and water. Which of the following foods would be easiest to digest?
  a. 10 g pasta
  b. 10 g strawberries
  c. 10 g slice of bread
  d. 10 g rice

  ANS: B DIF: III OBJ: 3-1.2

59. Our bodies are made of about 70 percent water. You would probably feel less thirsty if you would
  a. drink coffee.
  b. drink caffeinated soda.
  c. salt your food.
  d. eat fruits and vegetables that contain a lot of water.

  ANS: D DIF: II OBJ: 3-2.1

60. Kidney stones, caused by a buildup of salts and wastes in the kidneys,
  a. cause pain.
  b. usually pass naturally from the body.
  c. interfere with urine flow.
  d. All of the above

  ANS: D DIF: II OBJ: 3-2.3

COMPLETION

1. Urine travels from each kidney to the urinary bladder through the ____________________. (urethra or ureter)

  ANS: ureter DIF: I OBJ: 3-2.1

2. The rhythmic contractions that occur in the digestive tract are called ____________________. (peristalsis or enzymes)

  ANS: peristalsis DIF: I OBJ: 3-1.1

3. The chemical digestion of carbohydrates begins in the ____________________. (stomach or mouth)

  ANS: mouth DIF: I OBJ: 3-1.2

4. Bile is made in the ___________________ and stored in the ___________________. (liver/gallbladder or gallbladder/liver)

ANS: liver/gallbladder DIF: I OBJ: 3-1.1

5. ___________________ is the process of removing wastes and waste products from the body. This term is only used when substances are passed through a membrane before leaving the body. (Digestion or Excretion)

ANS: Excretion DIF: I OBJ: 3-2.1

6. Indigestible material is formed into feces in the ___________________. (large intestine or small intestine)

ANS: large intestine DIF: I OBJ: 3-1.1

7. The ___________________ system is a group of organs that work together to break down food. (urinary or digestive)

ANS: digestive DIF: I OBJ: 3-1.1

8. During swallowing, crushed food moves from the throat into the ___________________. (esophagus or small intestine)

ANS: esophagus DIF: I OBJ: 3-1.1

9. Bile is squeezed from the ___________________ into the small intestine where it breaks down fat droplets. (kidney or gallbladder)

ANS: gallbladder DIF: I OBJ: 3-1.1

10. The acid produced in the ___________________ kills most of the bacteria that are swallowed with food. (pancreas or stomach)

ANS: stomach DIF: I OBJ: 3-1.1

11. The ___________________ makes a juice that contains digestive enzymes, fats, and bicarbonates that neutralize the acid in chyme. (pancreas or liver)

ANS: pancreas DIF: I OBJ: 3-1.1

12. Enzymes chop up protein chains into ___________________.

ANS: amino acids DIF: I OBJ: 3-1.2

13. The outermost layer of a tooth, the ___________________, is the hardest material in the body.

ANS: enamel DIF: I OBJ: 3-1.1

14. As you chew, the food gets mixed with a liquid in your mouth called ____________________.

ANS: saliva DIF: I OBJ: 3-1.1

15. After a few hours of combined physical and chemical action in the stomach, your food will be reduced to a soupy mixture called ____________________.

ANS: chyme DIF: I OBJ: 3-1.2

16. The ____________________ is a fish-shaped organ located between the stomach and small intestine.

ANS: pancreas DIF: I OBJ: 3-1.1

17. Feces pass to the outside through an opening called the ____________________.

ANS: anus DIF: I OBJ: 3-1.1

18. The ____________________ stores feces until they can be expelled through the anus.

ANS: rectum DIF: I OBJ: 3-1.1

19. The stomach is blocked off at either end by bands of muscle called ____________________.

ANS: sphincters DIF: I OBJ: 3-1.1

20. An open sore in the stomach lining is called a ____________________.

ANS: gastric ulcer DIF: I OBJ: 3-1.3

21. The ____________________ system removes waste products from your blood.

ANS: urinary DIF: I OBJ: 3-2.1

22. ____________________ are the filters within each kidney.

ANS: Nephrons DIF: I OBJ: 3-2.1

23. The green fluid that is stored in the gallbladder is called ____________________.

ANS: bile DIF: I OBJ: 3-1.1

24. The yellow fluid that remains in the nephrons is called ____________________.

ANS: urine DIF: I OBJ: 3-2.2

25. Urine is stored in the ____________________.

ANS: urinary bladder DIF: I OBJ: 3-2.1

26. ____________________ is the process of expelling urine from the body.

ANS: Urination DIF: I OBJ: 3-2.1

27. The balance of fluids is controlled by chemical messengers in the body called ____________________.

ANS: hormones DIF: I OBJ: 3-2.1

28. Salts and wastes that collect inside the kidneys can form ____________________, which painfully interfere with urine flow.

ANS: kidney stones DIF: I OBJ: 3-2.3

SHORT ANSWER

1. What is the difference between mechanical digestion and chemical digestion?

ANS:
Mechanical digestion usually involves the crushing and mashing of whole bites of food. Mechanical digestion prepares the food for chemical digestion, which is the process of changing large molecules of food into smaller molecules that can be absorbed into the bloodstream.

DIF: I OBJ: 3-1.2

2. Give two reasons why the following statement is true: Digestion begins in the mouth.

ANS:
Mechanical and physical digestion begin in the mouth. As you chew, you are physically digesting your food. Saliva contains an enzyme that begins the chemical digestion of carbohydrates.

DIF: II OBJ: 3-1.2

3. What happens to the food that you eat when it gets to the stomach?

ANS:
The contractions of the stomach break food into smaller pieces. Chemical digestion by enzymes, water, and acids breaks bonds in the food.

DIF: I OBJ: 3-1.2

4. Describe the roles of the liver, the gallbladder, and the pancreas in digestion.

ANS:
The liver stores nutrients and makes bile, which is used in the digestion of fat. Bile is stored in a small, baglike organ called the gallbladder. The pancreas makes pancreatic juice, which contains enzymes for digesting proteins, starches, and fats and contains bicarbonate for neutralizing the acid in chyme.

DIF: I OBJ: 3-1.1

5. How would the inability to make saliva affect digestion?

ANS:
Without saliva, swallowing would be more difficult and the digestion of carbohydrates would not begin until the food reached the stomach.

DIF: II OBJ: 3-1.1

6. Put the following statements about the urinary system in their proper order:

_____ a. Water is absorbed back into blood vessels.

_____ b. Urine leaves the kidney through the ureter.

_____ c. A large artery brings blood into the kidney.

_____ d. Water and other small substances, including glucose and urea, leave the blood vessels and enter the nephron.

ANS:
c–1, d–2, a–3, b–4

DIF: I OBJ: 3-2.1

7. What is the main function of the urinary system?

ANS:
The main function of the urinary system is to remove certain waste products from the blood.

DIF: I OBJ: 3-2.1

8. Which contains more water, the blood going into the kidney or the blood leaving the kidney? Explain.

ANS:
There is more water in the blood as it enters the kidney. Though the kidney returns most of the water back to the blood vessels, some of the water leaves the bloodstream when the kidneys produce urine.

DIF: II OBJ: 3-2.2

9. Give two reasons why it is important that the pancreas releases bicarbonate into the small intestine.

ANS:
Bicarbonate neutralizes the acidic chyme coming in from the stomach. This allows pancreatic enzymes and enzymes made in the intestine to work. By neutralizing the acid, the bicarbonate also protects the lining of the intestine.

DIF: I OBJ: 3-1.1

10. How does the structure of the small intestine improve its nutrient absorption?

ANS:
The long length, folds, and villi increase the surface area that comes into contact with the food that is being digested in the small intestine.

DIF: I OBJ: 3-1.1

11. What is a diuretic?

ANS:
A diuretic is a substance that causes the kidneys to produce more urine.

DIF: I OBJ: 3-2.2

12. Use the following terms to create a concept map: *teeth, stomach, digestion, bile, saliva, mechanical digestion, gallbladder, chemical digestion.*

ANS:

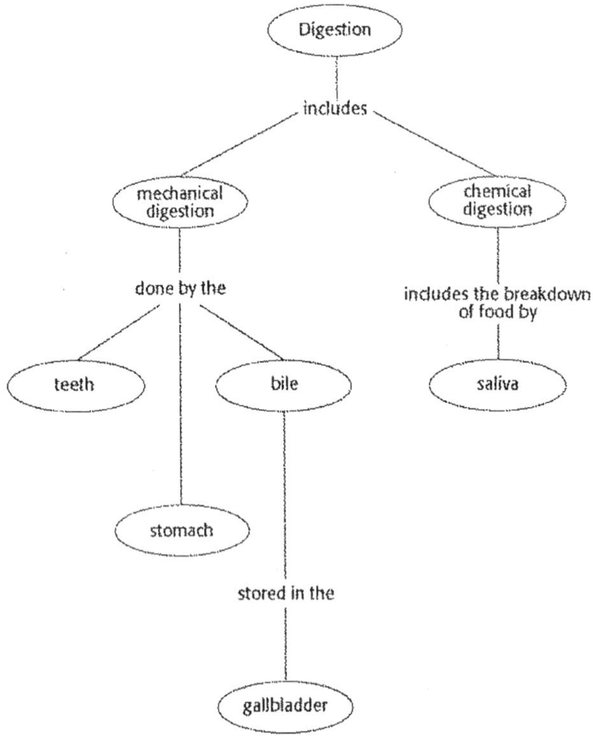

DIF: II OBJ: 3-1.1

13. How would digestion be affected if the liver were damaged?

ANS:
Sample answer: Digestion would be poor because the liver makes bile, which aids in the digestion of fats.

DIF: III OBJ: 3-2.3

14. Think about what happens when you put a piece of carbohydrate-dense food, such as bread, potato, or a cracker in your mouth. If you let a small piece sit near the tip of your tongue, it might begin to taste sweet. What digestive process would explain this change?

ANS:
Sample answer: Saliva causes chemical digestion of the carbohydrate-dense food, breaking it down to become simple sugars, which taste sweet.

DIF: II OBJ: 3-1.1

15. Mr. Jones has lost all of his molars and two of his premolars. How many teeth does Mr. Jones have?

ANS:
Mr. Jones has 18 remaining teeth.

DIF: II OBJ: 3-1.1

16. During a one-day water-balance study, a woman drank 1,500 mL of water. The food that she ate contained 750 mL of water, and 250 mL of water was produced internally during normal body processes. She lost 900 mL of water by sweating, 1,500 mL in urine, and 100 mL in feces. Overall, how much water did she gain or lose during the day?

ANS:
Overall, the woman gained no water and lost no water.

DIF: II OBJ: 3-2.1

The bar graph below shows how long the average meal spends in each portion of your digestive tract. Use this graph to answer the questions below.

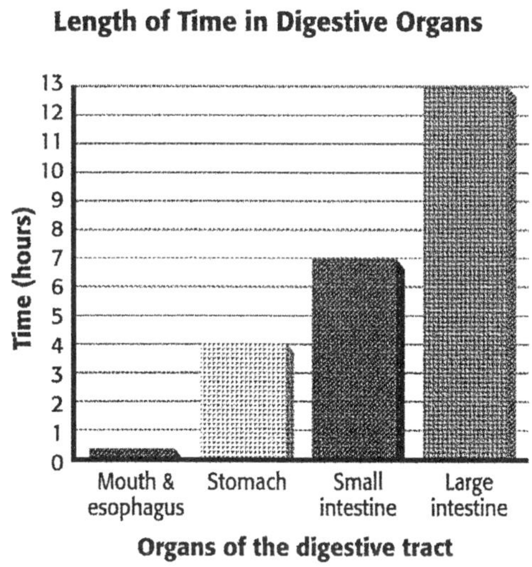

17. Where does the food spend the longest amount of time?

ANS:
Food spends the longest amount of time in the large intestine.

DIF: II OBJ: 3-1.1

18. On average, how much longer does food stay in the small intestine than in the stomach?

ANS:
On average food stays in the small intestine 3 hours longer than in the stomach.

DIF: II OBJ: 3-1.1

19. Which organ mixes food with special substances to make chyme? Approximately how long does food remain in this organ?

ANS:
The stomach mixes food with enzymes and acid to make a soupy mixture called chyme. Food stays in the stomach for approximately 4 hours.

DIF: II OBJ: 3-1.1

20. Bile breaks up large fat droplets into very small droplets. How long is the food in your body before it comes into contact with bile?

ANS:
Food is in your body just over 4 hours before it comes in contact with bile.

DIF: II OBJ: 3-1.1

21. What is the function of saliva?

ANS:
Saliva contains an enzyme that breaks down complex carbohydrates into simple sugars. This begins the process of chemical digestion. It also moistens food, making it easier to swallow.

DIF: I OBJ: 3-1.1

22. Describe two disorders of the urinary system.

ANS:
Kidney stones and bacterial infections are common disorders of the urinary system. Kidney stones are hardened salts and wastes that collect inside the kidneys. They can interfere with the flow of urine and can cause pain. Bacterial infections are painful infections that occur in the bladder and ureter. If bacterial infections are not treated, they can spread to the kidneys and cause permanent damage to the nephrons.

DIF: I OBJ: 3-2.3

23. Use the following terms to complete the concept map below: *urine, water, kidneys, ADH, urea, nephrons.*

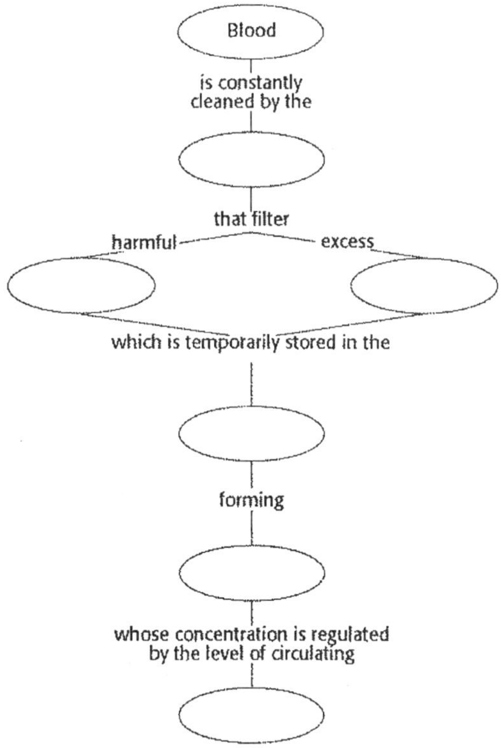

ANS:

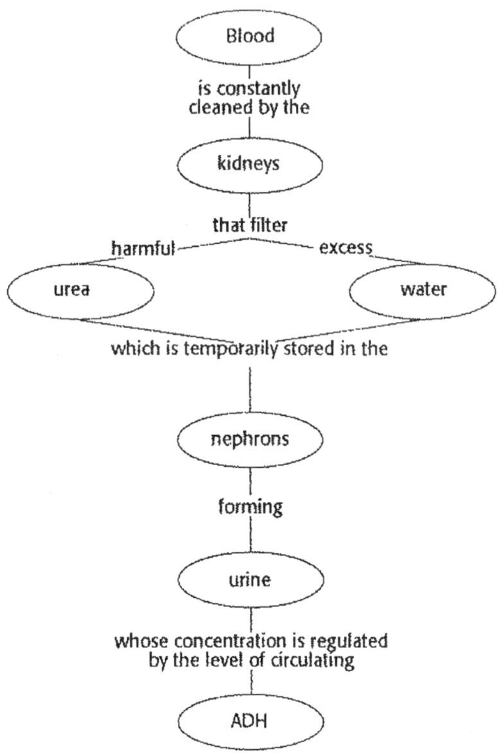

DIF: II OBJ: 3-2.2

24. Your kidneys filter about 2,000 L of blood each day. Approximately how many milliliters of blood do the kidneys filter each minute? Show your work.

ANS:
2,000 L / 24 h = 83.33 L/h
(83.33 L/h) / (60 min/hr) = 1.39 L/min
(1.39 L/min) x (1,000 mL/L) = 1,390 mL/min

DIF: II OBJ: 3-2.2

25. A few hours after drinking several caffeinated beverages, a man weighs himself. His weight has decreased by a small amount. Explain why this happened.

ANS:
Sample answer: His small weight loss was due to his high intake of caffeine. Caffeine is a diuretic, so it causes the kidneys to make more urine, decreasing the amount of water in the blood. The amount of fluid he lost in urine was greater than his fluid intake, so he lost a small amount of weight.

DIF: II OBJ: 3-2.1

Examine the diagram below, and answer the question that follows.

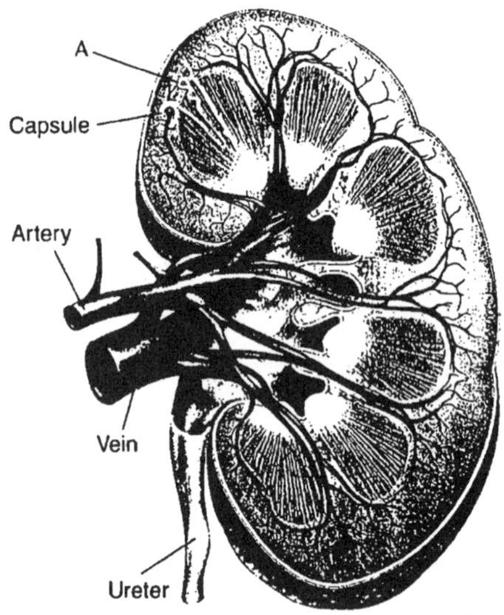

26. Give the name and function of structure **A**.

ANS:
A is a nephron. Nephrons are microscopic filters in the kidney. As blood passes through the kidneys, water and other substances are filtered into the nephron. Most of the water and some of the nutrients are reabsorbed into blood vessels and returned to circulate in the body. The remaining fluid becomes urine.

DIF: II OBJ: 3-2.2

27. What might happen if you were to swallow without having chewed your food thoroughly?

ANS:
Not chewing food thoroughly might slow digestion. Chewing food thoroughly does two things. First, it begins mechanical digestion by breaking, crushing, and mashing the food into smaller pieces. Second, chewing food thoroughly mixes the food with saliva, with begins chemical digestion.

DIF: II OBJ: 3-1.2

28. What is the advantage of chewing food *slowly*?

ANS:
Chewing food slowly gives the enzymes in your mouth more time to begin breaking down food before it enters the esophagus.

DIF: II OBJ: 3-1.2

29. You have probably noticed that the terms *urine, ureter, and urethra* all have a similar spelling. Distinguish between these three terms by explaining their role in excretion.

ANS:
Urine is a liquid waste in the body. The *ureter* and the *urethra* are tubes that help remove urine from the body.

DIF: II OBJ: 3-2.1

30. Explain the advantage of having mechanical digestion occur before and during chemical digestion.

ANS:
Mechanical digestion physically breaks down food by smashing it into smaller pieces. This increases the surface area of food upon which digestive enzymes can act during chemical digestion.

DIF: II OBJ: 3-1.2

31. Cellulose, a type of fiber found in fruits and vegetables, aids digestion by keeping some water in the contents of the large intestine. What could be the consequence of NOT eating enough fruits and vegetables?

ANS:
Not eating enough fruits and vegetables could slow down the movement of food through the large intestine. Not as much water would be retained by the contents of the large intestine, causing the material to dry out, possibly resulting in constipation.

DIF: II OBJ: 3-1.3

32. Diarrhea occurs when too little water is removed from digested food in the large intestine. Infer how diarrhea may cause dehydration, especially in infants and small children.

ANS:
The large intestine reabsorbs water from the undigested food. If diarrhea occurs, the body may lose more water than it takes in, and this leads to dehydration. Infants and small children have less water to lose, and dehydration can be dangerous to them.

DIF: III OBJ: 3-1.3

33. The liver captures and detoxifies many substances in the body. Predict how liver failure might affect the health of the rest of the body.

ANS:
The concentration of toxic substances in the body would increase, and eventually the toxic substances could cause death.

DIF: III OBJ: 3-1.3

34. How does a diuretic affect kidney function?

ANS:
Diuretics cause the kidneys to make more urine.

DIF: II OBJ: 3-2.2

35. Why is thirst often intensified by a diuretic?

ANS:
Thirst is often intensified by a diuretic because the diuretic causes the kidneys to make more urine, which decreases the amount of water in the blood. As the amount of water in the blood is decreased, your body asks for more water, and you feel more thirsty.

DIF: II OBJ: 3-2.2

36. If damage to nephrons prevents normal kidney functioning, what can be done to filter waste from the blood?

ANS:
A kidney machine, which pumps blood from an artery through a filter, can be used to remove waste from the blood and return cleaned blood to the body through a vein.

DIF: III OBJ: 3-2.3

TRUE/FALSE

1. Typically, a dendrite will transmit an electrical impulse to a neighboring axon.

   ANS: F DIF: I OBJ: 4-1.1

2. Most neurons are made up of either axons or dendrites; few neurons have both.

   ANS: F DIF: I OBJ: 4-1.1

3. The central nervous system is made up of the brain and the spinal cord.

   ANS: T DIF: I OBJ: 4-1.2

4. Rods typically function only in very bright light.

   ANS: F DIF: I OBJ: 4-2.2

5. Cones provide a colorful view of the world.

   ANS: T DIF: I OBJ: 4-2.2

6. Taste buds are located in tiny bumps on your tongue called papillae.

   ANS: T DIF: I OBJ: 4-2.3

7. The senses of smell and hearing are closely related. Without one, the other sense is dulled.

   ANS: F DIF: I OBJ: 4-2.3

8. The endocrine system is involved with high-speed processes.

   ANS: F DIF: I OBJ: 4-3.1

9. Hormones are chemicals secreted into the bloodstream.

   ANS: T DIF: I OBJ: 4-3.1

10. All endocrine glands come in pairs.

    ANS: F DIF: I OBJ: 4-3.2

11. The body uses feedback control to regulate the secretion of hormones.

    ANS: T DIF: I OBJ: 4-3.3

MULTIPLE CHOICE

1. Which of the following has receptors for smelling?
   a. cochlea cells
   b. thermoreceptors
   c. olfactory cells
   d. optic nerve

   ANS: C DIF: I OBJ: 4-2.3

2. Which of the following gives eyes their color?
   a. iris
   b. cornea
   c. lens
   d. retina

   ANS: A DIF: I OBJ: 4-2.2

3. Which of the following glands is associated with goiters?
   a. adrenal
   b. pituitary
   c. thyroid
   d. pancreas

   ANS: C DIF: II OBJ: 4-3.2

4. Which of the following is NOT part of the peripheral nervous system?
   a. spinal cord
   b. axons
   c. sensory receptors
   d. motor neurons

   ANS: A DIF: I OBJ: 4-1.2

5. Which part of the brain regulates blood pressure?
   a. right cerebral hemisphere
   b. left cerebral hemisphere
   c. cerebellum
   d. medulla

   ANS: D DIF: I OBJ: 4-1.3

6. Which of the following is associated with the endocrine system?
   a. reflex
   b. salivary gland
   c. fight-or-flight response
   d. voluntary response

   ANS: C DIF: I OBJ: 4-3.1

7. Long nerve fibers that convey electrical impulses to other cells are called
   a. axons.
   b. dendrites.
   c. organelles.
   d. neurons.

   ANS: A DIF: I OBJ: 4-1.1

8. Which of the following is NOT a function of the medulla?
   a. control of heart rate
   b. maintenance of blood pressure
   c. control of voluntary movement
   d. regulation of muscle tone in blood vessels

   ANS: C DIF: I OBJ: 4-1.3

9. Which of the following pairings of gland and hormone is NOT correct?
   a. pancreas : insulin
   b. adrenals : epinephrine
   c. thymus : glucagon
   d. pituitary : growth hormone

   ANS: C DIF: I OBJ: 4-3.2

10. The inability to produce insulin causes
   a. goiter.
   b. diabetes.
   c. fatigue.
   d. small stature.

   ANS: B DIF: I OBJ: 4-3.2

11. Which of the following allows you to smell?
   a. cochlea cells
   b. olfactory cells
   c. optic nerves
   d. cones

   ANS: B DIF: I OBJ: 4-2.3

12. Specialized cells called ____ transfer messages throughout your body in the form of fast-moving electrical energy.
   a. receptors
   b. hormones
   c. neurons
   d. All of the above

   ANS: C DIF: I OBJ: 4-1.1

13. A neuron consists of a
   a. nucleus, cell body, and DNA.
   b. cell body, axons, and impulses.
   c. nucleus, axons, dendrites.
   d. cell body, axons, dendrites.

   ANS: D DIF: I OBJ: 4-1.1

Examine the illustration of a neuron below and answer the questions that follow.

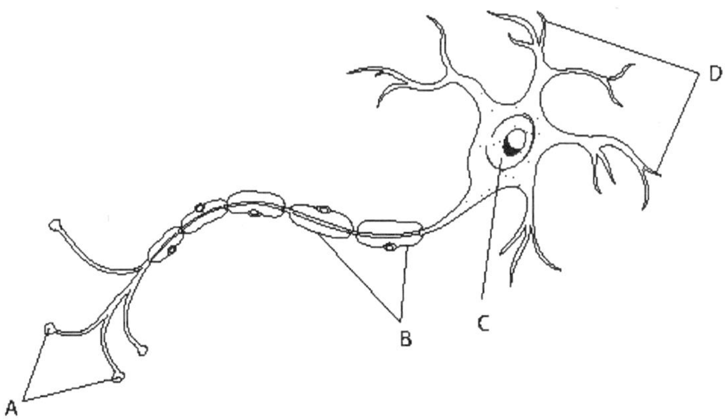

14. The structures labeled **A** are called
   a. dendrites.
   b. axon terminals.
   c. nodes.
   d. axons.

   ANS: B DIF: II OBJ: 4-1.1

15. The structures labeled **B** are called

a. dendrites.
b. nodes.
c. axon terminals.
d. axons.

ANS: D DIF: II OBJ: 4-1.1

16. The structure labeled **C** is called a(n)

a. axon.
b. nucleus.
c. cell body.
d. dendrite.

ANS: B DIF: II OBJ: 4-1.1

17. The structures labeled **D** are called

a. dendrites.
b. nuclei.
c. axon terminals.
d. axons.

ANS: A DIF: II OBJ: 4-1.1

18. From a neuron's cell body, information is transmitted to other cells by a fiber called a(n)

a. dendrite.
b. axon.
c. receptor.
d. lymphatic vessel.

ANS: B DIF: I OBJ: 4-1.1

19. Special neurons called ____ gather information about what is happening in and around your body.

a. sensory neurons
b. motor neurons
c. receptors
d. axons

ANS: A DIF: I OBJ: 4-1.1

20. Sensory neurons have special dendrites called ____ the detect changes inside and outside the body.

a. sensors
b. axon terminals
c. receptors
d. axons

ANS: C DIF: I OBJ: 4-1.1

21. Special neurons called ____ send impulses from the brain and spinal cord to other systems.

a. sensory neurons
b. motor neurons
c. receptors
d. axons

ANS: B DIF: I OBJ: 4-1.1

22. Which of the following would be responsible for letting your brain know when your stomach is full or empty?

a. motor neurons
b. receptors
c. axons
d. nerves

ANS: B DIF: I OBJ: 4-1.1

23. Which of the following would send messages to your sweat glands to release sweat?
   a. motor neurons
   b. sensory neurons
   c. axons
   d. nerves

   ANS: A   DIF: I   OBJ: 4-1.1

24. The nervous system's largest organ is the
   a. spinal cord.
   b. brain.
   c. axon that goes down your leg.
   d. skin.

   ANS: B   DIF: I   OBJ: 4-1.3

25. The largest part of your brain is called the
   a. medulla.
   b. cerebrum.
   c. cerebellum.
   d. pituitary.

   ANS: B   DIF: I   OBJ: 4-1.3

26. The second largest part of your brain is the
   a. cerebellum.
   b. pituitary.
   c. medulla.
   d. cerebrum.

   ANS: A   DIF: I   OBJ: 4-1.3

27. The part of your brain that connects to your spinal cord is called the
   a. cerebellum.
   b. cerebrum.
   c. medulla.
   d. pituitary.

   ANS: C   DIF: I   OBJ: 4-1.3

28. The part of your brain that receives sensory impulses from skeletal muscles and joints is the
   a. cerebellum.
   b. cerebrum.
   c. medulla.
   d. pituitary.

   ANS: A   DIF: I   OBJ: 4-1.3

29. Your ____ causes skeletal muscles to make adjustments in order to keep you upright.
   a. cerebellum
   b. medulla
   c. cerebrum
   d. pituitary

   ANS: A   DIF: I   OBJ: 4-1.3

30. Your ____ controls your heart rate, breathing, and other involuntary activities.
   a. pituitary
   b. cerebrum
   c. medulla
   d. cerebellum

   ANS: C   DIF: I   OBJ: 4-1.3

31. The ____ directs the left side of the body.
a. medulla
b. cerebellum
c. cerebrum's right hemisphere
d. cerebrum's left hemisphere

ANS: C DIF: I OBJ: 4-1.3

32. The ____ primarily controls activities such as speaking, reading, writing, and solving problems.
a. cerebellum
b. medulla
c. cerebrum's right hemisphere
d. cerebrum's left hemisphere

ANS: D DIF: I OBJ: 4-1.3

33. The ____ primarily controls activities that involve imagination, appreciation, and creativity.
a. medulla
b. cerebellum
c. cerebrum's right hemisphere
d. cerebrum's left hemisphere

ANS: C DIF: I OBJ: 4-1.3

34. Your ____ controls voluntary movements and allows you to detect touch, light, sound, odors, taste, pain, heat, and cold.
a. cerebellum
b. cerebrum
c. medulla
d. pituitary

ANS: B DIF: I OBJ: 4-1.3

35. A sensation that is detected by receptors in the skin is
a. deep pressure.
b. heat.
c. light touch.
d. All of the above

ANS: D DIF: I OBJ: 4-2.1

36. ____ is your awareness of light energy.
a. Touch
b. Taste
c. Vision
d. Smell

ANS: C DIF: I OBJ: 4-2.2

37. Visible light that is reflected by objects around you enters the eye through an opening at the front of your eye called the
a. retina.
b. optic nerve.
c. pupil.
d. photoreceptor.

ANS: C DIF: I OBJ: 4-2.2

38. The retina is packed with special neurons called
a. cornea.
b. pupils.
c. lenses.
d. photoreceptors.

ANS: D DIF: I OBJ: 4-2.3

39. Two types of photoreceptors in the eye are

a. rods and squares.
b. cones and rods.
c. ovals and cones.
d. squares and ovals.

ANS: B DIF: I OBJ: 4-2.3

40. ____ detect color.

a. Rods
b. Cones
c. Ovals
d. Squares

ANS: B DIF: I OBJ: 4-2.3

41. Nerve impulses that are detected by photoreceptors travel along axons and leave the back of each eye through a(n)

a. retina.
b. optic nerve.
c. pupil.
d. photoreceptor.

ANS: B DIF: I OBJ: 4-2.3

42. Cats are nocturnal animals, and they have adapted so that they can see very well in dim light. Therefore, cats may have more ____ than humans.

a. ovals
b. squares
c. rods
d. cones

ANS: C DIF: II OBJ: 4-2.3

43. Lemons look yellow because they reflect ____ light.

a. orange
b. red
c. yellow
d. green

ANS: C DIF: I OBJ: 4-2.3

44. In bright light, your pupil is

a. small.
b. large.
c. brown.
d. white.

ANS: A DIF: I OBJ: 4-2.2

45. When you look at objects close to the eye, the lens

a. gets flatter.
b. becomes more curved.
c. does not change, only the pupil does.
d. does not change, only the iris does.

ANS: B DIF: II OBJ: 4-2.2

46. When you look at objects far away, the lens

a. gets flatter.
b. becomes more curved.
c. does not change, only the pupil does.
d. does not change, only the iris does.

ANS: A DIF: II OBJ: 4-2.2

47. Nearsightedness occurs when the lens focuses the light
   a. in front of the cornea.
   b. in front of the pupil.
   c. in front of the retina.
   d. behind the retina.

   ANS: C DIF: II OBJ: 4-2.2

48. Farsightedness occurs when the lens focuses the light
   a. in front of the cornea.
   b. in front of the pupil.
   c. in front of the retina.
   d. behind the retina.

   ANS: D DIF: II OBJ: 4-2.2

49. A ____ is a group of cells that makes special chemicals for your body.
   a. hormone
   b. neuron
   c. dendrite
   d. gland

   ANS: D DIF: I OBJ: 4-3.1

50. The chemicals that are produced by the endocrine glands, or ____, are released into the bloodstream and carried to other places in the body.
   a. hormones
   b. neurons
   c. dendrites
   d. axons

   ANS: A DIF: I OBJ: 4-3.1

51. When you are frightened or angry, your ____ produce epinephrine, or adrenalin, which speeds up your heartbeat and breathing rate to prepare your body either to run from danger or to fight for survival.
   a. pituitary glands
   b. ovaries or testes
   c. adrenal glands
   d. parathyroid glands

   ANS: C DIF: I OBJ: 4-3.2

52. Your ____ stimulates skeletal growth, helps the thyroid function properly, regulates the amount of water in the blood, and stimulates the birth process in pregnant women.
   a. pituitary gland
   b. pancreas
   c. ovaries
   d. testes

   ANS: A DIF: II OBJ: 4-3.2

53. Your ____ increases the rate at which you use energy.
   a. pituitary gland
   b. thyroid gland
   c. thymus gland
   d. parathyroid gland

   ANS: B DIF: II OBJ: 4-3.2

54. Your ____ regulates the immune system, which helps your body fight disease.
   a. adrenal gland
   b. parathyroid gland
   c. thymus gland
   d. thyroid gland

   ANS: C DIF: II OBJ: 4-3.2

55. Suppose you have a friend that eats right and exercises regularly, but she is still overweight and has trouble losing it. It is possible that your friend might have a problem with her
a. adrenal gland.
b. thyroid gland.
c. thymus gland.
d. pancreas.

ANS: B DIF: III OBJ: 4-3.2

56. Some people develop a condition called *diabetes mellitus*, which is the inability for the ____ to make enough insulin to keep his or her blood sugar levels within safe limits.
a. adrenal gland
b. thyroid gland
c. thymus gland
d. pancreas

ANS: D DIF: II OBJ: 4-3.2

57. In females, the ____ produce hormones involved in reproduction.
a. adrenal glands
b. testes
c. ovaries
d. pancreas

ANS: C DIF: II OBJ: 4-3.2

58. If an immune system has difficulty receiving signals to fight a disease, the ____ might be affected.
a. thymus gland
b. adrenal gland
c. parathyroid gland
d. thyroid gland

ANS: A DIF: II OBJ: 4-3.2

59. After you eat a meal, the bloodstream absorbs ____ from the small intestine.
a. hormones
b. glucose
c. neurons
d. receptors

ANS: B DIF: I OBJ: 4-3.3

60. When the glucose level in the blood is high, the pancreas
a. absorbs glucose from the blood.
b. releases glucose into the blood.
c. absorbs insulin from the blood.
d. releases insulin into the blood.

ANS: D DIF: I OBJ: 4-3.3

61. Insulin signals the ____ to take in glucose from the body, convert it to glycogen, and store it for future energy needs.
a. pancreas
b. liver
c. cerebrum
d. cerebellum

ANS: B DIF: I OBJ: 4-3.3

62. When the blood sugar level returns to normal, the ____ stops releasing insulin.
a. pancreas
b. liver
c. cerebrum
d. cerebellum

ANS: A DIF: I OBJ: 4-3.3

63. To keep your blood sugar level from falling below normal, you must
a. receive a shot of insulin.
b. drink plenty of water.
c. eat.
d. avoid eating sugary foods.

ANS: C DIF: I OBJ: 4-3.3

64. Sometimes a child's ____ does not make enough growth hormone, causing the child's growth to be stunted.
a. thymus gland
b. adrenal gland
c. pituitary gland
d. thyroid gland

ANS: C DIF: II OBJ: 4-3.2

65. If you touch a hot burner on a stove, your hand suddenly jerks away. This involuntary action is
a. a reflex.
b. an impulse.
c. an overreaction.
d. a receptor.

ANS: A DIF: I OBJ: 4-1.3

66. Suppose you are on a game show that requires you to press a buzzer with your right hand when you know the answer. The part of the brain that controls the pressing of the buzzer is the
a. medulla.
b. cerebellum.
c. cerebrum's right hemisphere.
d. cerebrum's left hemisphere.

ANS: D DIF: II OBJ: 4-1.3

67. Composing music on the guitar involves using the
a. medulla.
b. cerebellum.
c. cerebrum's right hemisphere.
d. cerebrum's left hemisphere.

ANS: C DIF: II OBJ: 4-1.3

68. Competing in a spelling bee involves the
a. medulla.
b. cerebellum.
c. cerebrum's right hemisphere.
d. cerebrum's left hemisphere.

ANS: D DIF: II OBJ: 4-1.3

69. Some figure skaters can balance on one foot while gliding over the ice. This action involves their
a. medulla.
b. cerebellum.
c. pancreas.
d. pituitary gland.

ANS: B DIF: II OBJ: 4-1.3

COMPLETION

1. Your brain and spinal cord make up your ______________________. (central nervous system or peripheral nervous system)

ANS: central nervous system DIF: I OBJ: 4-1.2

2. Sensory receptors in the ____________________ detect vibrations. (cochlea or eardrum)

ANS: cochlea DIF: I OBJ: 4-2.3

3. Epinephrine is produced by the adrenal glands in response to ____________________. (glucose or stress)

ANS: stress DIF: I OBJ: 4-3.2

4. The part of a neuron that passes an impulse to other cells is the ____________________. (dendrite or axon terminal)

ANS: axon terminal DIF: I OBJ: 4-1.1

5. The medulla is mostly responsible for activities that are ____________________. (involuntary or voluntary)

ANS: involuntary DIF: I OBJ: 4-1.3

6. Receptors that can covert light into impulses are found in the ____________________. (olfactory cells or retina)

ANS: retina DIF: I OBJ: 4-2.2

7. The brain sends messages through ____________________ to skeletal muscles, causing them to contract. (sensory neurons or motor neurons)

ANS: motor neurons DIF: I OBJ: 4-1.2

8. ____________________ are the short, branched extensions of a nerve-cell body that receive information. (Dendrites or Axons)

ANS: Dendrites DIF: I OBJ: 4-1.2

9. ____________________ are photoreceptors that can detect very dim light. (Cones or Rods)

ANS: Rods DIF: I OBJ: 4-2.2

10. Most memories are stored in the ____________________. (cerebrum or cerebellum)

ANS: cerebrum DIF: I OBJ: 4-1.3

11. The light-sensitive area in back of the eye is called the ____________________. (iris or retina)

ANS: retina DIF: I OBJ: 4-2.2

12. Motor neurons are part of the ____________________ nervous system. (central or peripheral)

ANS: peripheral DIF: I OBJ: 4-1.2

13. The ____________________ nervous system consists of communication pathways, or nerves, that connect all areas of your body to your central nervous system.

ANS: peripheral DIF: I OBJ: 4-1.2

14. The electrical messages that pass along the neurons are called ____________________.

ANS: impulses DIF: I OBJ: 4-1.1

15. ____, which extend throughout your body, are axons bundled together with blood vessels and connective tissue.

ANS: Nerves DIF: I OBJ: 4-1.1

16. Many of the processes that the brain controls that happen automatically are referred to as ____________________.

ANS: involuntary DIF: I OBJ: 4-1.3

17. Your ____________________, part of the central nervous system, is about as big around as your thumb and contains neurons and bundles of axons that pass impulses to and from the brain.

ANS: spinal cord DIF: I OBJ: 4-1.3

18. The spinal cord is surrounded by protective bones called ____________________.

ANS: vertebrae DIF: I OBJ: 4-1.3

19. A quick, involuntary action that helps protect your body from damage is called a ____________________.

ANS: reflex DIF: I OBJ: 4-1.3

20. A ring of muscle fibers allows the ____________________ to open and close, making the pupil change size.

ANS: iris DIF: I OBJ: 4-2.2

21. A ____________________ is a piece of curved material behind the pupil that allows light to pass through but changes its direction so that is focused on the retina.

ANS: lens DIF: I OBJ: 4-2.2

22. The ____________________ is a tiny snail-shaped organ of the inner ear.

ANS: cochlea DIF: I OBJ: 4-2.3

23. The tongue is covered with tiny bumps called ____________________ within which the taste buds are embedded.

ANS: papillae DIF: I OBJ: 4-2.3

24. The receptors for smell are located on ____________________ cells in the upper part of your nasal cavity.

ANS: olfactory DIF: I OBJ: 4-2.3

25. Your ____________________ system is involved with the control of slower, long-term processes such as fluid balance, growth, and sexual development.

ANS: endocrine DIF: I OBJ: 4-3.1

26. Endocrine glands know when to start and stop hormone release because your body has special systems called ________________________.

ANS: feedback controls DIF: I OBJ: 4-3.3

27. When sound waves reach the outer ear, they are funneled into the middle ear, where they cause the ____________________ to vibrate.

ANS: eardrum DIF: I OBJ: 4-2.3

28. It is in your ____________________ that you have thoughts, feelings, and memories about sensations.

ANS: brain DIF: II OBJ: 4-1.3

SHORT ANSWER

1. Make a labeled diagram that shows the path of an electrical message from one neuron to another neuron.

   ANS:
   Diagrams will vary but should show the following path: dendrite, cell body, axon, axon terminal, dendrite.

   DIF: I OBJ: 4-1.1

2. Explain how the peripheral nervous system connects with the central nervous system.

   ANS:
   The peripheral nervous system consists of communication pathways, or nerves, that connect all areas of your body to your central nervous system.

   DIF: I OBJ: 4-1.2

3. If a spider is crawling up your left arm, which cerebral hemisphere controls the movement that you will use to knock it off?

ANS:
You would use the left hemisphere, because you would have to use your right hand to knock it off.

DIF: I OBJ: 4-1.3

4. List the three major parts of the brain, and describe their functions.

ANS:
The three major parts of the brain are the cerebrum, the cerebellum, and the medulla. Functions of the cerebrum include thinking, storing memories, controlling voluntary movements, and detecting sensations such as light and sound. The cerebellum monitors the position of the body and helps to maintain balance.The medulla controls involuntary processes such as blood pressure, heart rate, and involuntary breathing.

DIF: I OBJ: 4-1.3

5. Describe a time when you experienced a reflex.

ANS:
Answers will vary but should demonstrate an understanding that reflexes involve involuntary muscle actions.

DIF: II OBJ: 4-1.3

6. List three sensations that receptors in the skin can detect.

ANS:
Answers should include three of the following: pain, heat, pressure, cold, and touch.

DIF: I OBJ: 4-2.1

7. Explain why you would have trouble seeing bright colors at a candlelit dinner.

ANS:
Cones, which perceive color, do not function well in dim light. Rods function in dim light but only perceive tones of gray.

DIF: I OBJ: 4-2.2

8. How is your sense of taste similar to your sense of smell?

ANS:
Receptors for smell detect chemicals that are present in air. Receptors for taste detect chemicals that are present in your mouth.

DIF: I OBJ: 4-2.3

9. If you can focus on objects close to you but things become blurry when they are far away, would a concave or a convex lens correct your vision?

ANS:
A concave lens would correct my vision.

DIF: II OBJ: 4-2.2

10. What is the function of the endocrine system?

ANS:
Your endocrine system controls body processes such as fluid balance, growth, and sexual development with the use of chemicals called hormones.

DIF: I OBJ: 4-3.1

11. Why are feedback controls important?

ANS:
Feedback controls stop and start hormone release.

DIF: I OBJ: 4-3.3

12. Name four endocrine glands, and tell what each one does in the body.

ANS:
Answers will vary but should include four of the following glands: thyroid gland, parathyroid glands, thymus gland, ovaries, pituitary gland, adrenal glands, pancreas, and testes.

DIF: I OBJ: 4-3.2

13. Epinephrine, the fight-or-flight hormone, increases the level of glucose in the blood. Why would this be important in times of stress?

ANS:
During stressful situations, your body functions might need to speed up, requiring more energy.

DIF: II OBJ: 4-3.2

14. Look around the classroom for an example of a feedback control. Draw a diagram explaining how this feedback control works to start and stop an action.

ANS:
Answers will vary, but students might draw a diagram demonstrating how a thermostat keeps water or air within a certain temperature range.

DIF: II OBJ: 4-3.3

15. Describe several situations in which your adrenal glands might release epinephrine, causing you to have a fight-or-flight reaction.

ANS:
Answers will vary but should describe dangerous or stressful situations that the student has encountered.

DIF: I OBJ: 4-3.1

16. What causes the size of your pupils to change?

ANS:
Light stimulates a ring of muscle fibers, which enable the iris to open and close.

DIF: I OBJ: 4-2.2

17. a. What is a reflex?
b. How does a reflex enable you to act quickly?

ANS:
a. A reflex is a quick, involuntary action.
b. Because the impulse does not need to travel to the brain before a muscle can respond, the involuntary reaction can be very quick.

DIF: I OBJ: 4-1.3

18. What is the function of the middle-ear bones?

ANS:
The vibrations of the eardrum make the ear bones vibrate, one of which vibrates against the cochlea, creating waves that are detected by neurons.

DIF: I OBJ: 4-2.3

19. Using the terms you learned in this chapter, write down a step-by-step sequence for the path taken by an impulse, beginning at a pain receptor in your left big toe. Be sure to mention each kind of neuron and its parts as well as specific organs in the nervous system.

ANS:
The impulse travels from sensory receptors in the toe along axons leading to the spinal cord. Motor neurons carry a new message to the muscles in the leg telling them to move the foot. The pain message continues along the spinal cord to the cerebrum.

DIF: I OBJ: 4-1.1

20. Use the following terms to create a concept map: *the nervous system, spinal cord, medulla, peripheral nervous system, brain, cerebrum, central nervous system, cerebellum.*

ANS:

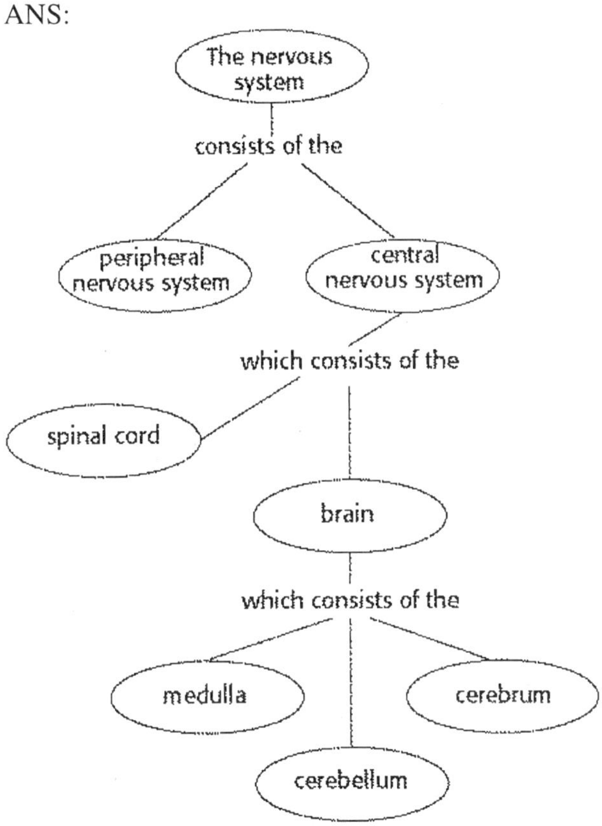

DIF: II OBJ: 4-1.2

21. Why is it important to have a lens that can change shape inside the eye?

ANS:
The lens must change shape to focus objects that are varying distances from the eye.

DIF: II OBJ: 4-2.2

22. Why can the nervous system have a faster effect on the body than the endocrine system?

ANS:
Electrical messages travel along nerves faster than chemicals can travel in the bloodstream.

DIF: II OBJ: 4-3.1

23. Why is it important that reflexes occur without thought?

ANS:
The speed of a reflex might prevent serious harm to the body.

DIF: II OBJ: 4-1.3

24. Sound travels about 335 m/s (1 km is equal to 1,000 m). How many kilometers would a sound travel in 1 minute?

ANS:
Sound would travel 20.1 km in 1 minute.

DIF: II OBJ: 4-2.3

25. Some axons can send one impulse every 0.4 milliseconds. One second is equal to 1,000 milliseconds. How many impulses could one of these axons send every second?

ANS:
One of these axons could send 2,500 impulses/second.

DIF: II OBJ: 4-1.1

Look at the drawing below, and answer the questions that follow.

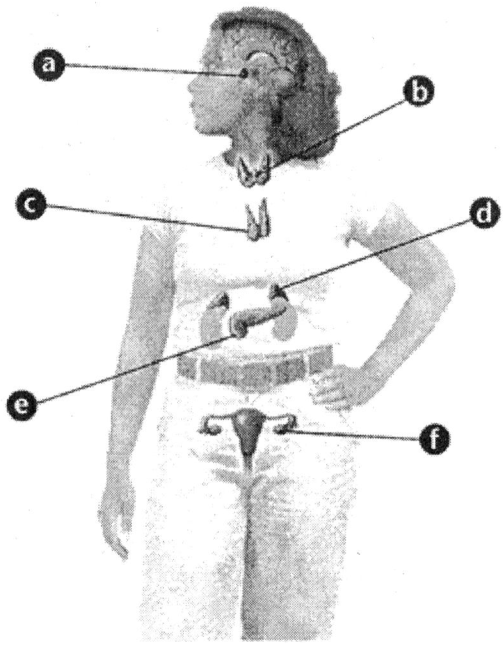

26. Which letter identifies the gland that regulates blood sugar?

ANS:
The letter *e* identifies the pancreas, which regulates blood sugar.

DIF: II OBJ: 4-3.2

27. Which letter identifies the gland that releases a hormone that stimulates the birth process in pregnant women?

ANS:
The letter *a* identifies the pituitary gland, which stimulates the birth process in pregnant women.

DIF: II OBJ: 4-3.2

28. Which letter identifies the gland that helps the body fight disease?

ANS:
The letter *c* identifies the thymus, which helps the body fight disease.

DIF: II OBJ: 4-3.2

29. Explain how the role of the nervous system differs from that of the endocrine system.

ANS:
The main role of the nervous system is the control of rapid, short-term movements or actions. The endocrine system is responsible for slower, long-term processes, such as fluid balance, growth, sexual development, and reproductive function.

DIF: I OBJ: 4-3.1

30. Explain what happens within your nervous system when you step on a tack with your bare foot.

ANS:
When you step on a tack, a message of "pain" travels to your spinal cord, and a message to move your leg travels back to your leg muscles immediately. This is known as a reflex. Your brain does not receive the message until after your foot has moved.

DIF: I OBJ: 4-1.3

31. The surface of your tongue is covered with about 10,000 taste buds. Suppose the surface area of your tongue is 21 cm^2. Calculate the average number of taste buds per square millimeter. Show your work.

ANS:
1 $cm^2 = (10\ mm)^2 = 100\ mm^2$. The surface area of the tongue in millimeters is
21 $cm^2 \times 100\ mm^2/cm^2 = 2{,}100\ mm^2$. Therefore, 10,000 taste buds/2,100 mm^2 = 4.76 taste buds per square millimeter.

DIF: I OBJ: 4-2.3

32. Why would an injury to the medulla be more life threatening than an injury to the cerebrum?

ANS:
The cerebrum is the area of the brain where thinking is carried out. For example, processes such as reading, writing, and problem-solving are controlled by the cerebrum. These processes are not necessary for survival. However, basic life functions, such as heart rate, blood pressure, involuntary respiration, and body temperature are controlled by the medulla. Major damage to the medulla can result in death.

DIF: I OBJ: 4-1.3

The two diagrams below show a human eye focusing on different objects. Examine the diagrams and answer the questions that follow.

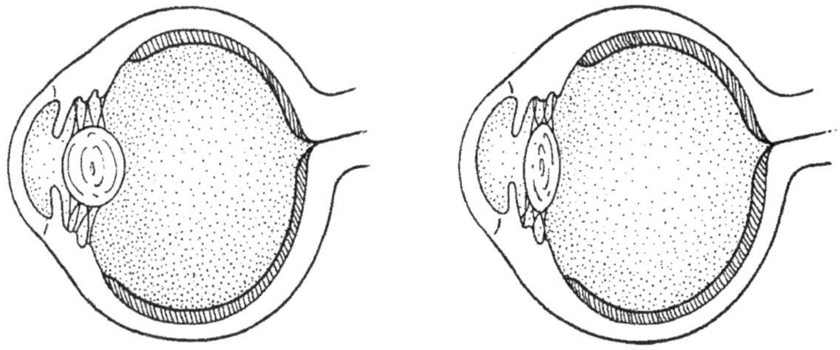

A B

33. What structure in the eye is changing shape between *A* and *B*?

ANS:
The structure changing shape is the lens.

DIF: II OBJ: 4-2.2

34. Which diagram represents an eye looking at a nearby object? Explain.

ANS:
A represents an eye looking at a nearby object because the lens of the eye is thicker and more curved.

DIF: II OBJ: 4-2.2

35. Use the following terms to complete the concept map below: *chemical stimuli, taste buds, cochlea, papillae, electrical impulses, olfactory cells.*

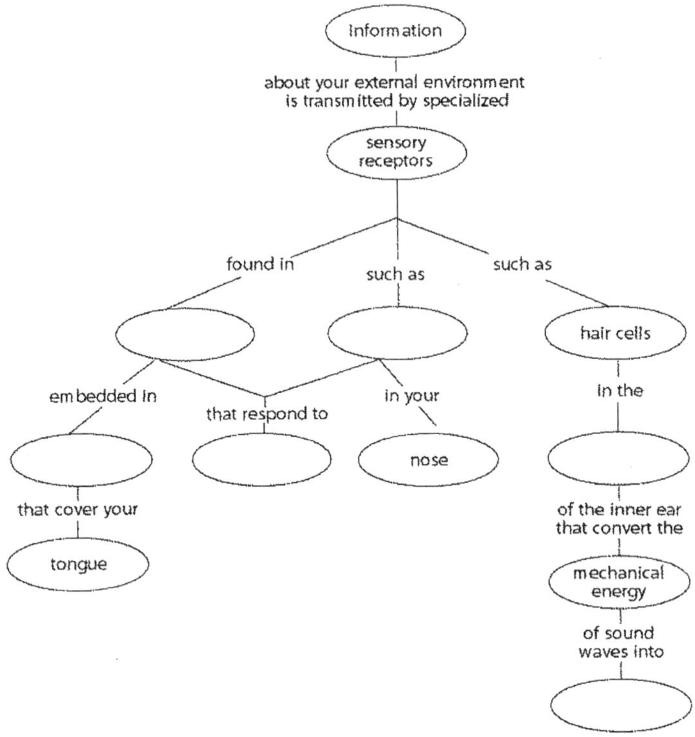

ANS:

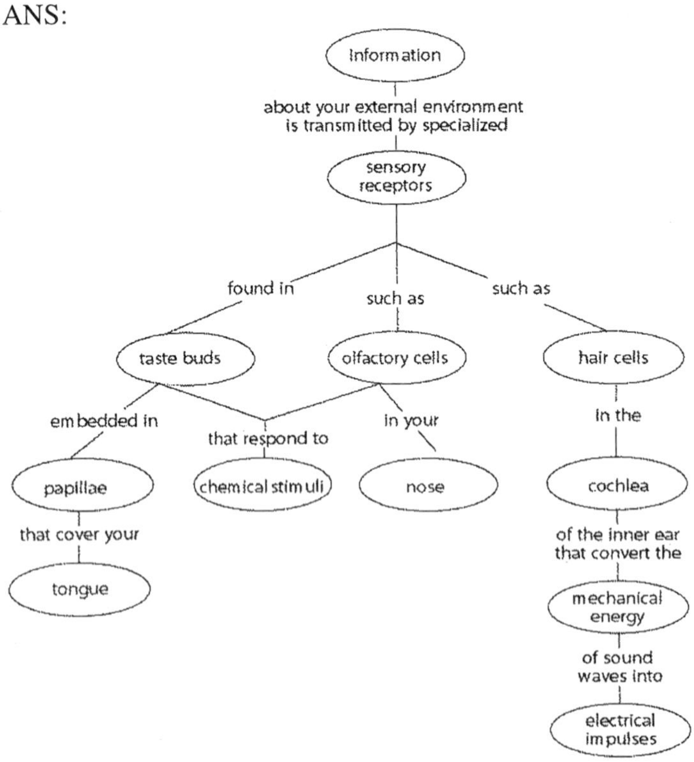

DIF: II OBJ: 4-2.3

36. What is the advantage of having a ring of muscle fibers that causes the iris to open and close and allows the pupil change size?

ANS:
The ring of muscle fibers regulates the amount of light that passes to the retina, which helps you see in both bright light and dim light.

DIF: I OBJ: 4-2.2

37. Suppose a seven-year-old boy is much taller than would be expected for his age. Explain what gland might have caused this to happen and why.

ANS:
The pituitary gland could have produced too much growth hormone at an early age, which is why the boy is much taller than expected.

DIF: II OBJ: 4-3.3

38. What may happen if the spinal cord is injured?

ANS:
If the spinal cord is injured, any sensory information coming into it from below where the damage occurred may be unable to travel to the brain. Likewise, any motor commands the brain sends to an area below the injury may not get through to the peripheral nerves.

DIF: II OBJ: 4-1.3

MULTIPLE CHOICE

1. The sea star can reproduce asexually by
   a. fragmentation.
   b. budding.
   c. external fertilization.
   d. internal fertilization.

   ANS: A DIF: I OBJ: 5-1.1

2. The correct path of sperm through the male reproductive system is
   a. testes → epididymis→ urethra → vas deferens.
   b. epididymis → urethra → testes → vas deferens.
   c. testes → vas deferens → epididymis → urethra.
   d. testes → epididymis → vas deferens → urethra.

   ANS: D DIF: I OBJ: 5-3.1

3. If the first day of the menstrual cycle is the onset of menstruation, on what day does ovulation typically occur?
   a. 2nd day
   b. 5th day
   c. 14th day
   d. 28th day

   ANS: C DIF: I OBJ: 5-2.1

4. Monotremes are different from placental mammals because they
   a. are mammals.
   b. have hair.
   c. nurture their young with milk.
   d. lay eggs.

   ANS: D DIF: I OBJ: 5-1.3

5. All of the following are sexually transmitted diseases EXCEPT
   a. chlamydia.
   b. AIDS.
   c. infertility.
   d. genital herpes.

   ANS: C DIF: I OBJ: 5-2.2

6. Fertilization occurs in the ____, and implantation occurs in the ____.
   a. uterus, fallopian tube
   b. fallopian tube, vagina
   c. uterus, vagina
   d. fallopian tube, uterus

   ANS: D DIF: I OBJ: 5-3.1

7. After the eighth week, the developing human is called a(n)
   a. zygote.
   b. embryo.
   c. fetus.
   d. infant.

   ANS: C DIF: I OBJ: 5-3.2

8. The process of ovulation in humans
   a. begins at puberty.
   b. occurs several times a month.
   c. moves an egg from the uterus to the ovary.
   d. removes an unfertilized egg from the female's body.

   ANS: A DIF: I OBJ: 5-2.1

9. Infertility can be caused by all of the following EXCEPT
   a. a low sperm count.
   b. inability to ovulate.
   c. ectopic pregnancies.
   d. sexually transmitted diseases.

   ANS: C DIF: I OBJ: 5-2.2

10. Around the beginning of the fourth week, a developing human has
    a. nostrils.
    b. a beating heart.
    c. fingers and toes.
    d. eyelids that flutter.

    ANS: B DIF: I OBJ: 5-3.2

11. Sperm are produced in the
    a. vas deferens.
    b. epididymis.
    c. seminiferous tubules.
    d. prostate gland.

    ANS: C DIF: I OBJ: 5-2.1

12. As soon as menstruation is over, the uterus prepares for
    a. implantation.
    b. budding.
    c. ovulation.
    d. fragmentation.

    ANS: C DIF: I OBJ: 5-2.1

13. ____ is an asexual form of reproducing that occurs when a small part of the parent's body develops into an independent organism.
    a. Budding
    b. Rooting
    c. Fragmentation
    d. Splitting

    ANS: A DIF: I OBJ: 5-1.1

14. ____ is an asexual form of reproducing that occurs when an organism breaks into two or more parts, each of which may grow into a separate individual.
    a. Budding
    b. Fragmentation
    c. Rooting
    d. Splitting

    ANS: B DIF: I OBJ: 5-1.1

15. In sexual reproduction, the female parent produces sex cells called
    a. sperm.
    b. zygote.
    c. eggs.
    d. chromosomes.

    ANS: C DIF: I OBJ: 5-1.1

16. The male parent produces sex cells called
    a. sperm.
    b. chromosomes.
    c. eggs.
    d. zygote.

    ANS: A DIF: I OBJ: 5-1.1

17. When an egg's nucleus joins with a sperm's nucleus, a new kind of cell is created, called a(n)
    a. egg.
    b. zygote.
    c. sperm.
    d. chromosome.

    ANS: B DIF: I OBJ: 5-1.1

18. All human cells—except egg and sperm cells—contain ____ chromosomes.
    a. 22
    b. 23
    c. 46
    d. 48

    ANS: C DIF: I OBJ: 5-1.1

19. Eggs and sperm EACH contain ____ chromosomes.
    a. 22
    b. 23
    c. 46
    d. 48

    ANS: B DIF: I OBJ: 5-1.1

20. Eggs and sperm are formed by a process called
    a. budding.
    b. fragmentation.
    c. meiosis.
    d. mitosis.

    ANS: C DIF: I OBJ: 5-1.1

21. In humans, meiosis involves the division of one cell with 46 chromosomes into four sex cells with ____ chromosomes each.
    a. 11 or 12
    b. 23
    c. 46
    d. 184

    ANS: B DIF: II OBJ: 5-1.1

22. When an egg and sperm join to form a zygote, the zygote contains ____ chromosomes.
    a. 11 or 12
    b. 23
    c. 46
    d. 184

    ANS: C DIF: II OBJ: 5-1.1

23. External fertilization takes place in a ____ environment.
    a. moist
    b. dry
    c. warm
    d. cold

    ANS: A DIF: I OBJ: 5-1.2

24. Which of the following animals reproduces by external fertilization?
   a. a bird
   b. a frog
   c. a lizard
   d. a horse

   ANS: B DIF: II OBJ: 5-1.2

25. Which of the following animals reproduces asexually?
   a. a fish
   b. a frog
   c. a sea star
   d. a platypus

   ANS: C DIF: II OBJ: 5-1.1

26. Which of the following animals lays fertilized eggs?
   a. a frog
   b. a snake
   c. a cat
   d. a sea star

   ANS: B DIF: II OBJ: 5-1.1

27. Mammals that lay eggs are
   a. monotremes.
   b. duotremes.
   c. placental mammals.
   d. marsupials.

   ANS: A DIF: I OBJ: 5-1.3

28. Mammals that give birth to live young that are only partially developed are
   a. monotremes.
   b. duotremes.
   c. placental mammals.
   d. marsupials.

   ANS: D DIF: I OBJ: 5-1.3

29. ____ nourish their young internally before birth.
   a. Monotremes
   b. Duotremes
   c. Placental mammals
   d. Marsupials

   ANS: C DIF: I OBJ: 5-1.3

30. ____ with pouches have extra bones to help support the weight of their young.
   a. Monotremes
   b. Duotremes
   c. Placental mammals
   d. Marsupials

   ANS: D DIF: I OBJ: 5-1.3

31. The echidna and the platypus are mammals that lay eggs. They are two families of
   a. monotremes.
   b. duotremes.
   c. placental mammals.
   d. marsupials.

   ANS: A DIF: I OBJ: 5-1.3

32. The ____ make sperm and testosterone.
a. testes
b. penis
c. epididymis
d. vas deferens

ANS: A DIF: I OBJ: 5-2.1

33. The principal male sex hormone is
a. estrogen.
b. progesterone.
c. testosterone.
d. adrenalin.

ANS: C DIF: I OBJ: 5-2.1

34. ____ is a hormone that regulates the production of sex cells and the development of male characteristics.
a. Adrenalin
b. Progesterone
c. Estrogen
d. Testosterone

ANS: D DIF: I OBJ: 5-2.1

35. Which statement below best describes the reason that the two testes rest in the scrotum instead of within the rest of the body?
a. If the testes were within the body, they would interfere with urination.
b. There is no other place in the body to put the testes.
c. Testes must be surrounded by fat cells, and the scrotum is the only place to put fat without damaging vital organs.
d. Sperm cannot develop properly at the high temperatures within the human body.

ANS: D DIF: I OBJ: 5-2.1

36. A skin-covered sac that hangs from a male body is called the
a. scrotum.
b. sphincter.
c. anus.
d. rectum.

ANS: A DIF: I OBJ: 5-2.1

37. The scrotum is about ____ the body.
a. the same temperature as
b. two degrees cooler than
c. two degrees warmer than
d. four degrees warmer than

ANS: B DIF: I OBJ: 5-2.1

38. A healthy adult male produces several hundred million ____ each day.
a. testis
b. ovaries
c. eggs
d. sperm

ANS: D DIF: I OBJ: 5-2.1

39. ____ is the time of life when the sex organs of both males and females become mature.
a. Implantation
b. Childhood
c. Puberty
d. Adulthood

ANS: C DIF: I OBJ: 5-2.1

40. Before sperm leave a testis, they are stored in a tube called a(n)

a. vas deferens.
b. epididymis.
c. urethra.
d. penis.

ANS: B DIF: I OBJ: 5-2.1

41. The ____ connect each epididymis to the body.

a. vas deferens
b. seminiferous tubules
c. urethra
d. penis

ANS: A DIF: I OBJ: 5-2.1

42. As sperm swim through the vas deferens, they mix with fluids from several glands, creating a mixture called

a. testosterone.
b. estrogen.
c. semen.
d. mucus.

ANS: C DIF: I OBJ: 5-2.1

43. To leave the body, semen passes through the vas deferens into the ____, a tube that runs through the penis.

a. seminiferous tubules
b. urethra
c. ovaries
d. fallopian tube

ANS: B DIF: I OBJ: 5-2.1

44. The ____ transfers semen into the female's body during sexual intercourse.

a. penis
b. scrotum
c. testis
d. vas deferens

ANS: A DIF: I OBJ: 5-3.1

45. In the female reproductive system, the ____ produce(s) the eggs.

a. uterus
b. ovaries
c. fallopian tubes
d. vagina

ANS: B DIF: I OBJ: 5-2.1

46. The two ovaries produce sex hormones, such as ____, that regulate the release of eggs and direct the development of female characteristics.

a. estrogen and testosterone
b. estrogen and progesterone
c. testosterone and progesterone
d. adrenalin and testosterone

ANS: B DIF: I OBJ: 5-2.1

47. An ovary contains

a. only one egg at a time.
b. many eggs, each at the same stage of development.
c. many eggs at different stages of development.
d. the uterus and the vagina.

ANS: C DIF: I OBJ: 5-2.1

48. During ____, an egg is ejected through the ovary wall.

a. puberty
b. menstruation
c. implantation
d. ovulation

ANS: D DIF: I OBJ: 5-2.1

49. The ____ leads from each ovary to the uterus.

a. vas deferens
b. fallopian tube
c. vagina
d. urethra

ANS: B DIF: I OBJ: 5-2.1

50. The ____ is the organ where a baby grows and develops.

a. uterus
b. fallopian tube
c. vagina
d. urethra

ANS: A DIF: I OBJ: 5-2.1

51. Every month starting at ____, the lining of the uterus thickens in preparation for pregnancy.

a. puberty
b. menstruation
c. implantation
d. ovulation

ANS: A DIF: I OBJ: 5-2.1

52. ____ is the monthly discharge of blood and tissue from the uterus.

a. Puberty
b. Menstruation
c. Implantation
d. Ovulation

ANS: B DIF: I OBJ: 5-2.1

53. The menstrual cycle usually occurs every ____ days, and lasts about 5 days.

a. 7
b. 14
c. 28
d. 90

ANS: C DIF: II OBJ: 5-2.1

54. What happens to an egg if it is NOT fertilized by the time it reaches the uterus?

a. It turns into a zygote.
b. It moves back into an ovary.
c. It is removed through menstruation.
d. It stays in the vagina until it is fertilized.

ANS: C DIF: I OBJ: 5-2.1

55. In the United States, about 15 percent of married couples cannot have children, meaning they

a. have gonorrhea.
b. have chlamydia.
c. have syphilis.
d. are infertile.

ANS: D DIF: I OBJ: 5-2.2

56. The natural process of creating a human baby starts when
a. a man deposits millions of sperm into a woman's vagina.
b. an embryo is formed.
c. a fetus is formed.
d. ovulation occurs.

ANS: A DIF: I OBJ: 5-3.1

57. During sexual intercourse, most of the sperm will die because the vagina is
a. very hot.
b. very cold.
c. acidic.
d. alkaline

ANS: C DIF: I OBJ: 5-3.1

58. The egg is fertilized in the
a. uterus.
b. fallopian tube.
c. vagina.
d. ovaries.

ANS: B DIF: I OBJ: 5-3.1

59. The surviving sperm cover the egg, releasing ____ that help dissolve the egg's outer covering.
a. enzymes
b. hormones
c. chromosomes
d. DNA

ANS: A DIF: I OBJ: 5-3.1

60. As soon as one sperm gets through, a membrane closes around the fertilized egg, which
a. prevents other sperm from entering.
b. encourages other sperm to enter.
c. drops it into the ovaries.
d. Both (b) and (c)

ANS: A DIF: I OBJ: 5-3.1

61. It takes about ____ for a fertilized egg to travel down a woman's fallopian tube and reach her uterus.
a. two hours
b. five days
c. two weeks
d. nine months

ANS: B DIF: I OBJ: 5-3.1

62. By the time a fertilized egg reaches the uterus, it is a tiny ball of cells called a(n)
a. zygote.
b. embryo.
c. fetus.
d. child.

ANS: B DIF: I OBJ: 5-3.1

63. A woman is officially pregnant when
a. the egg is fertilized.
b. a zygote is formed.
c. a fetus is formed.
d. the embryo implants itself in her uterus.

ANS: D DIF: II OBJ: 5-3.1

64. The ____ contains a network of blood vessels that provide the embryo with oxygen and nutrients from the mother's blood.
    a. uterus
    b. umbilical cord
    c. placenta
    d. amnion

    ANS: C DIF: I OBJ: 5-3.2

65. The thin, fluid-filled membrane surrounding the embryo is called the
    a. uterus.
    b. umbilical cord.
    c. placenta.
    d. amnion.

    ANS: D DIF: I OBJ: 5-3.2

66. The ____ connects the embryo to the placenta.
    a. vas deferens
    b. umbilical cord
    c. fallopian tube
    d. amnion

    ANS: B DIF: I OBJ: 5-3.2

67. Human infancy lasts about
    a. two months.
    b. eight months.
    c. two years.
    d. five years.

    ANS: C DIF: II OBJ: 5-3.2

68. Human childhood extends over
    a. a few weeks.
    b. a few months.
    c. a few years.
    d. a decade.

    ANS: D DIF: I OBJ: 5-3.2

COMPLETION

1. Reptiles, birds, and mammals reproduce sexually by ________________________. (internal fertilization or external fertilization)

   ANS: internal fertilization DIF: I OBJ: 5-1.2

2. Sperm are produced in the ________________________ within the testes. (epididymis or seminiferous tubules)

   ANS: seminiferous tubules DIF: I OBJ: 5-2.1

3. The sperm-containing fluid that exits the male body is known as ____________________. (semen or amniotic fluid)

   ANS: semen DIF: I OBJ: 5-2.1

4. The release of an egg from the ovary occurs once each month and is called ____________________. (ovulation or menstruation)

ANS: ovulation DIF: I OBJ: 5-2.1

5. The organ of exchange between the developing embryo and the mother is the ____________________. (amnion or placenta)

ANS: placenta DIF: I OBJ: 5-2.1

6. A ____________________ is the cell created when a sperm fertilizes an egg. (fetus or zygote)

ANS: zygote DIF: I OBJ: 5-2.1

7. The ____________________ is the organ where a mammalian fetus grows and develops. (uterus or vagina)

ANS: uterus DIF: I OBJ: 5-2.1

8. The ____________________ is the common channel through which urine and sperm exit the human male. (vas deferens or urethra)

ANS: urethra DIF: I OBJ: 5-2.1

9. Reproduction in frogs and toads usually occurs by ____________________ fertilization. (external or internal)

ANS: external DIF: I OBJ: 5-1.2

10. ____________________ reproduction results in offspring that are genetically identical to the parent. (Sexual or Asexual)

ANS: Asexual DIF: I OBJ: 5-1.1

11. When ____________________ eggs hatch, the newborn young are very undeveloped. (monotreme or marsupial)

ANS: monotreme DIF: I OBJ: 5-1.3

12. In ____________________ reproduction, a single parent has offspring that are genetically identical to itself.

ANS: asexual DIF: I OBJ: 5-1.1

13. Genes are located in ____________________.

ANS: chromosomes DIF: I OBJ: 5-1.1

14. ____________________ fertilization occurs when the sperm fertilizes the eggs outside the female's body.

ANS: External DIF: I OBJ: 5-1.2

15. ____________________ fertilization occurs when the eggs and sperm join inside the female's body.

ANS: Internal DIF: I OBJ: 5-1.2

16. ____________________ twins result from one egg dividing into two eggs after becoming fertilized by the same sperm.

ANS: Identical DIF: II OBJ: 5-2.2

17. ____________________ twins result when two eggs are released by an ovary and are both fertilized by different sperm.

ANS: Fraternal DIF: II OBJ: 5-2.2

18. In an ____________________ pregnancy, the fertilized egg attaches itself to a fallopian tube or another area of the reproductive system.

ANS: ectopic DIF: I OBJ: 5-2.2

19. ____ are diseases that can pass from an infected person to an uninfected person during sexual contact.

ANS: Sexually transmitted diseases (STDs)

DIF: I OBJ: 5-2.2

20. ____________________, the uncontrolled division of cells, sometimes occurs in the reproductive organs.

ANS: Cancer DIF: I OBJ: 5-2.2

21. Only 30 percent of all embryos successfully embed themselves in the thick, nutrient-rich lining of their mother's uterus, a process called ____________________.

ANS: implantation DIF: I OBJ: 5-3.1

22. At eight weeks old, the embryo is developed enough to be called a ____________________.

ANS: fetus DIF: I OBJ: 5-3.2

23. During ____________________, the fetus is usually squeezed headfirst through the vagina.

ANS: labor DIF: I OBJ: 5-3.2

24. The baby's navel is all that remains of the point where the ________________________ was attached.

ANS: umbilical cord DIF: I OBJ: 5-3.2

25. During birth, the baby is usually pushed headfirst through the ___________________.

ANS: vagina DIF: I OBJ: 5-3.2

SHORT ANSWER

1. How many parents are needed to reproduce asexually?

ANS:
One parent is needed.

DIF: I OBJ: 5-1.1

2. What is the difference between monotremes and marsupials?

ANS:
Monotremes lay eggs and marsupials give birth to live young.

DIF: I OBJ: 5-1.3

3. How is a zygote formed?

ANS:
A zygote is formed when a sperm enters an egg and the nuclei of the egg and sperm join.

DIF: I OBJ: 5-1.2

4. Birds lay eggs, but they are not considered monotremes. Explain why.

ANS:
Birds are not mammals. They do not nurture their young with milk, and they are not covered with hair.

DIF: II OBJ: 5-1.3

5. What is the difference between sperm and semen?

ANS:
A sperm is a male sex cell, while semen is the fluid that contains sperm.

DIF: I OBJ: 5-2.1

6. Can a woman become pregnant at any time of the month? Explain.

ANS:
No; usually only one egg per month is released, and it is viable for only a few days after ovulation.

DIF: I OBJ: 5-2.1

7. Define *sexually transmitted diseases*, and give three examples.

ANS:
Sexually transmitted diseases are diseases passed through sexual contact. Some examples are chlamydia, gonorrhea, genital herpes, and AIDS. (Students must list three.)

DIF: I OBJ: 5-2.2

8. a. How are the ovaries similar to the testes?
   b. How are the ovaries different from the testes?

ANS:
a. The ovaries and the testes release sex cells and hormones.
b. The testes continuously produce masses of sex cells, but the ovaries release only one mature sex cell per month.

DIF: II OBJ: 5-2.1

9. What is the difference between an embryo and a fetus?

ANS:
An embryo is less developed than a fetus.

DIF: I OBJ: 5-3.1

10. Why does a membrane form around an egg once a sperm has entered?

ANS:
The membrane keeps other sperm from entering. This is important because it helps maintain the characteristic human chromosome number.

DIF: I OBJ: 5-3.1

11. What developmental changes take place from birth to puberty?

ANS:
When a person is first born, he or she is an infant. During infancy, the teeth and nervous system develop. During childhood, beginning at age 2, physical and mental capabilities develop. During adolescence, sexual characteristics develop.

DIF: I OBJ: 5-3.2

12. When astronauts work in space, they are sometimes attached to the spacecraft by a line called an umbilical. Why do you think the line has been given this name?

ANS:
The astronaut's umbilical cord supplies life support for the astronaut just as a fetus' umbilical cord provides life support for the fetus.

DIF: II OBJ: 5-3.2

13. How do the offspring created by asexual reproduction differ from those created by sexual reproduction?

ANS:
Offspring from asexual reproduction are genetically identical to the parent, while offspring from sexual reproduction have genetic material from two parents.

DIF: I OBJ: 5-1.1

14. How do internal and external fertilization differ? How are they alike?

ANS:
Internal fertilization occurs inside the female's body, while external fertilization occurs outside the female's body. Both result in the creation of single cells with combined genetic information from the parents.

DIF: I OBJ: 5-1.2

15. What is puberty?

ANS:
Puberty is the time of life when sex organs become mature.

DIF: I OBJ: 5-2.1

16. What purpose does the epididymis serve?

ANS:
It serves as a temporary storage site for sperm before they leave the testes.

DIF: I OBJ: 5-2.1

17. What is menstruation?

ANS:
Menstruation is the monthly discharge of blood and tissue from the uterus.

DIF: I OBJ: 5-2.1

18. What is implantation?

ANS:
It is the process by which an embryo embeds itself in the uterus.

DIF: I OBJ: 5-3.1

19. What functions does the placenta serve?

ANS:
It is a two-way exchange organ that allows oxygen and nutrients to travel to the fetus from the mother and allows wastes to travel from the fetus to the mother.

DIF: I OBJ: 5-3.2

20. a. What human reproductive organs produce sperm?
b. What human reproductive organs produce egg cells?

ANS:
a. The testes produce sperm.
b. The ovaries produce eggs.

DIF: I OBJ: 5-2.1

21. Through what structure does oxygen from the mother pass into the fetus' body?

ANS:
Oxygen passes from the mother's body to the fetus' body through the placenta.

DIF: I OBJ: 5-3.2

22. What are four stages of human life following birth?

ANS:
Students' answers should include four of the following: infancy, childhood, adolescence, young adulthood, middle-age adulthood, or older adulthood.

DIF: I OBJ: 5-3.2

23. What two cells combine to make a zygote?

ANS:
An egg and a sperm combine to form a zygote.

DIF: I OBJ: 5-2.1

24. What is the difference between budding and fragmentation?

ANS:
Budding is a type of asexual reproduction in which a young organism develops off a small part of the parent. Fragmentation is a type of asexual reproduction in which an organism develops from a separated piece of the parent.

DIF: I OBJ: 5-1.1

25. Use the following terms to create a concept map: *asexual reproduction, budding, external fertilization, fragmentation, reproduction, internal fertilization, sexual reproduction.*

ANS:

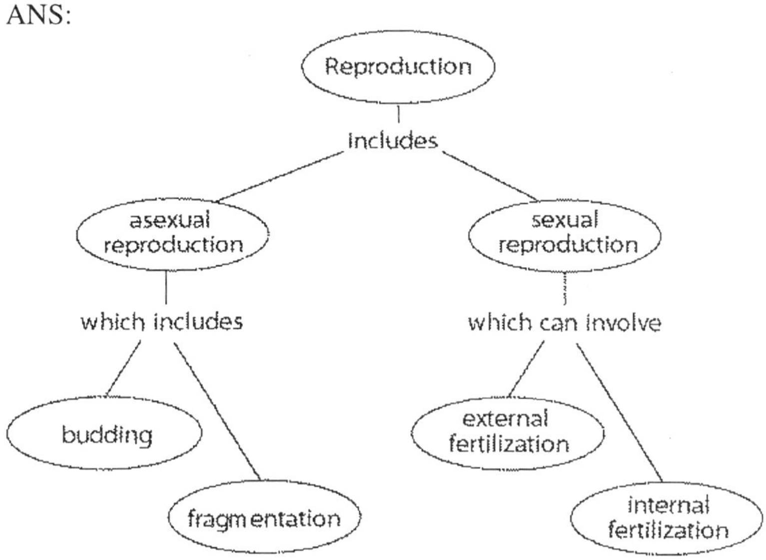

DIF: II OBJ: 5-1.2

26. Explain why the testes are found in the scrotum instead of inside the male body.

ANS:
Sperm develop best at a temperature three degrees lower than normal body temperature. Therefore, the testes are suspended away from the body to keep them cooler.

DIF: II OBJ: 5-2.1

27. a. What is the function of the uterus?
b. How is the function of the uterus related to the menstrual cycle?

ANS:

a. The uterus is the organ in the female reproductive system in which an embryo can develop into a fetus. Every month it builds up tissue that can help nourish a developing embryo.
b. If no embryo is present, the tissue will discharge and cause menstruation.

DIF: II OBJ: 5-2.1

28. How is meiosis important to human reproduction?

ANS:
Meiosis ensures that the characteristic human chromosome number remains the same. Since two cells must combine to form a zygote, each sex cell needs only half the 46 chromosomes that all other human body cells have.

DIF: II OBJ: 5-2.1

29. Hardy Junior High School has 2,750 students. If 1 pair of identical twins is born for every 250 births, about how many pairs of identical twins will be attending the school?

ANS:
There would be about 11 pairs of identical twins in the school.

DIF: II OBJ: 5-2.2

30. Mrs. Schmidt had a baby April 30th. Her baby developed inside her uterus for 9 months. What month was her egg fertilized?

ANS:
The egg was fertilized in August.

DIF: II OBJ: 5-3.2

31. In the United States, seven infants die before their first birthday for every 1,000 births. Convert this figure to a percentage. Is your answer greater than or less than 1%?

ANS:
Approximately 0.7 percent of American infants die before their first birthday. This is less than 1 percent.

DIF: II OBJ: 5-3.2

32. In Haiti, a small country in the Caribbean, 74 infants die before their first birthday for every 1,000 births.
   a. Convert this figure to a percentage. Is your answer greater or less than 1 percent?
   b. Why do you think there is such a difference between the United States and Haiti?

ANS:

   a. Approximately 7.4 percent of Haitian infants die before their first birthday. This is more than 1 percent.
   b. The Haitian infant mortality rate is higher than the American rate because Americans generally have access to better health care than Haitians.

DIF: II OBJ: 5-3.2

Look at the following graph, which illustrates the cycles of the male hormone, testosterone, and the female hormone, estrogen, over a period of 28 days. Use the graph to answer the questions that follow.

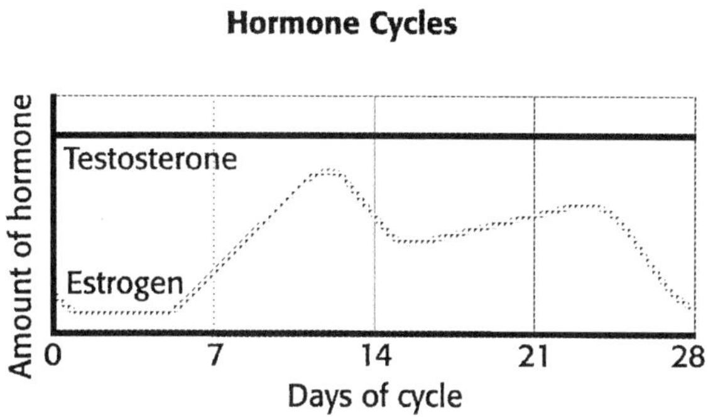

33. What is the major difference between the two hormone levels over the 28-day period?

ANS:
Estrogen levels fluctuate, but testosterone levels stay at the same level throughout the month.

DIF: II OBJ: 5-2.1

34. What cycle do you think estrogen affects?

ANS:
Estrogen affects the menstrual cycle.

DIF: II OBJ: 5-2.1

35. Why might the level of testosterone stay the same?

ANS:
Testosterone levels stay the same because men continually produce sperm.

DIF: II OBJ: 5-2.1

36. How is budding different from fragmentation?

ANS:
Budding occurs when a small part of the parent's body develops into a new individual. Fragmentation occurs when an organism breaks into two or more parts and each part grows into a separate individual.

DIF: I OBJ: 5-1.1

37. What characteristics make marsupials different from placental mammals?

ANS:
Marsupials give birth to live young that are only partially developed. Most marsupials have pouches where their young develop. When a baby marsupial first attaches itself to its mother's nipple, the nipple expands to prevent the baby from separating. Placental mammals retain their young inside the uterus until they are further along in development.

DIF: I OBJ: 5-1.3

38. A newborn's skull is made up of bone plates connected by soft, flexible material. This material allows a newborn's skull to change shape. Why is this characteristic important?

ANS:
This characteristic is important because it allows the newborn to be squeezed through the narrow pelvic opening without damage to the skull during birth.

DIF: II OBJ: 5-3.2

39. Use the following terms to complete the concept map below: *fallopian tubes, epididymis, uterus, testes, ovaries, seminiferous tubules.*

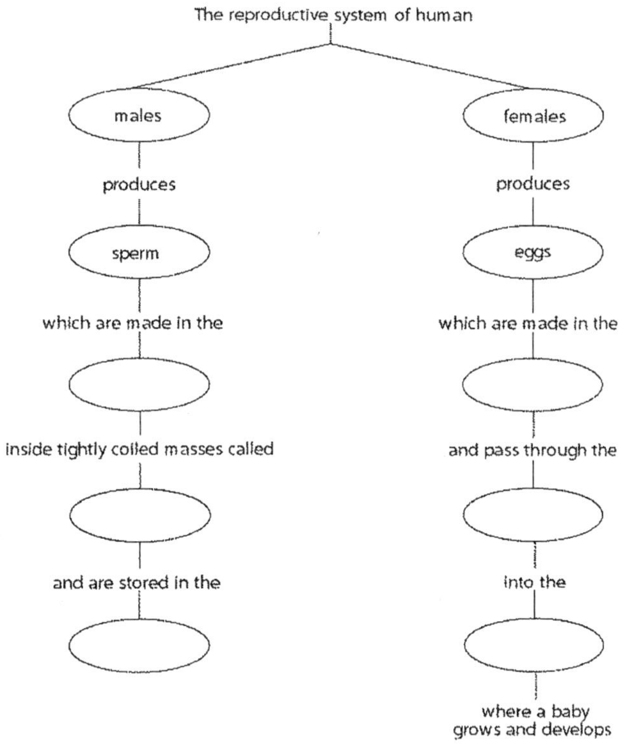

ANS:

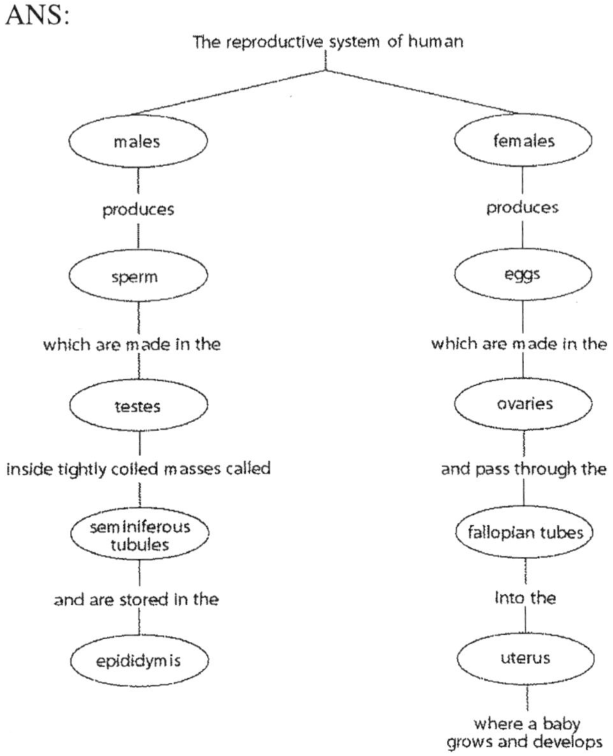

DIF: II OBJ: 5-2.1

Below is a diagram of a fully developed human fetus. Study the diagram and answer the questions that follow.

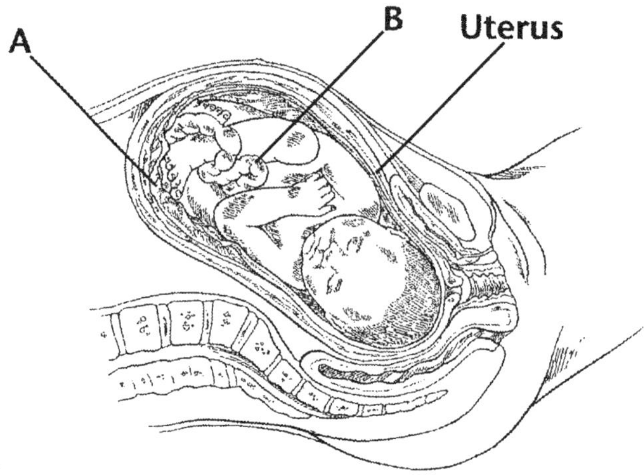

40. a. What does **A** represent?
    b. What is the function of this organ?

ANS:
a. **A** represents the placenta.
b. The placenta provides a developing baby with oxygen and nutrients from the mother's blood. It also removes waste materials from the embryo or fetus and delivers the waste to the mother for excretion.

DIF: II OBJ: 5-3.2

41. a. What does **B** represent?
    b. What is the function of this structure?

ANS:
a. **B** represents the umbilical cord.
b. The umbilical cord connects the developing fetus to the placenta allowing blood to flow between the mother and the fetus.

DIF: II OBJ: 5-3.2

42. What is the difference between asexual and sexual reproduction?

ANS:
In *asexual reproduction*, one animal produces offspring that are genetically identical to itself. In *sexual reproduction,* the genes of two individuals are mixed when sex cells join to form a zygote. This zygote develops into a unique individual.

DIF: I OBJ: 5-1.1

43. Suppose a town has 10,000 married couples. If 15 percent of married couples are infertile, how many couples in the town would be infertile?

ANS:
$10{,}000 \times 0.15 = 1{,}500$
About 1,500 couples in the town would be infertile.

DIF: II OBJ: 5-2.2

44. Siamese twins are twins that are born with their bodies joined together in some way. From what you know about how twins are formed, explain how this could happen.

ANS:
Siamese twins are produced when a fertilized egg does not separate completely into two halves. The zygotes successfully implanted themselves in the uterine wall, increasing the likelihood that they would be born joined together.

DIF: III OBJ: 5-2.2

45. The word menstruation comes from the Greek word *menos*, meaning "month." When a woman reaches the later stages of adulthood, usually between the ages of 40 or 50, she goes through *menopause*. Infer what stage of life menopause might be for a woman.

ANS:
The prefix, *men-*, means "month." Pause means "to stop." Perhaps menopause is a time in a woman's life when her monthly cycle stops, or she stops menstruating.

DIF: III OBJ: 5-3.2

46. Some infertile couples can take advantage of biotechnological advances such as *in vitro fertilization* to enable them to have a baby. *In vitro* means "in glass." The woman who undergoes in vitro fertilization is first given a fertility drug or hormone to stimulate her ovaries to release more than one egg at a time. Then, the eggs are removed and placed in a glass Petri dish. A drop of her husband's sperm is added. After many cell divisions have taken place, the doctor inserts the ball of cells into the woman's uterus. Explain the process of in vitro fertilization using the terms *zygote, implantation, and embryo.*

ANS:
A *zygote* is formed *in vitro* when sperm combines with the egg. After a period of cell division, the doctor places the *embryo* in the uterus. The embryo undergoes *implantation* when it embeds itself in the walls of the uterus.

DIF: III OBJ: 5-3.1

TRUE/FALSE

1. All diseases are caused by pathogens.

   ANS: F DIF: I OBJ: 6-1.1

2. You can become infected with some pathogens by shaking hands with an infected person.

   ANS: T DIF: I OBJ: 6-1.2

3. Leaving food out at room temperature helps kill bacteria.

   ANS: F DIF: I OBJ: 6-1.2

4. Vaccines kill or inhibit the growth of bacteria.

   ANS: F DIF: I OBJ: 6-1.3

5. Pasteurization uses very cold temperatures to kill bacteria.

   ANS: F DIF: I OBJ: 6-1.3

6. As dead skin cells drop off your body, some pathogens may enter your skin.

   ANS: F DIF: I OBJ: 6-2.1

7. Oil secreted by oil glands in your skin have chemicals that kill pathogens.

   ANS: T DIF: I OBJ: 6-2.1

8. Antibodies are specific to certain pathogens.

   ANS: T DIF: I OBJ: 6-2.2

9. Fevers are dangerous because they can increase the growth of pathogens in your body and make you sicker.

   ANS: F DIF: II OBJ: 6-2.3

10. An allergy is caused by the immune system attacking its own body's cells.

    ANS: F DIF: I OBJ: 6-3.1

11. Killer T cells are useful at killing cancer cells.

    ANS: T DIF: I OBJ: 6-3.2

12. *HIV* and *AIDS* are two terms for the same disease.

ANS: F DIF: II OBJ: 6-3.3

13. People with AIDS have too many killer T cells. These cells begin to attack the body's cells, and this results in AIDS.

ANS: F DIF: II OBJ: 6-3.3

14. The AIDS virus causes HIV.

ANS: F DIF: II OBJ: 6-3.3

MULTIPLE CHOICE

1. Pathogens are
   a. all viruses and microorganisms.
   b. viruses and microorganisms that cause disease.
   c. noninfectious organisms.
   d. all bacteria that live in water.

ANS: B DIF: I OBJ: 6-1.1

2. The following is an infectious disease:
   a. allergies
   b. rheumatoid arthritis
   c. asthma
   d. common cold

ANS: D DIF: I OBJ: 6-1.1

3. The skin keeps pathogens out by
   a. staying warm enough to kill pathogens.
   b. releasing killer T cells onto the surface.
   c. shedding dead cells and secreting oils.
   d. All of the above

ANS: C DIF: I OBJ: 6-2.1

4. Memory B cells
   a. kill pathogens.
   b. activate killer T cells.
   c. activate killer B cells.
   d. produce B cells that make antibodies.

ANS: D DIF: I OBJ: 6-2.2

5. A fever
   a. slows pathogen growth.
   b. helps B cells multiply faster.
   c. helps T cells multiply faster.
   d. All of the above

ANS: D DIF: I OBJ: 6-2.3

6. Macrophages
   a. make antibodies.
   b. release helper T cells.
   c. live in the gut.
   d. engulf pathogens.

   ANS: D DIF: I OBJ: 6-2.2

7. Which of the following is NOT a way your body protects itself against disease?
   a. acids in your throat
   b. enzymes in your eyes, nose, and mouth
   c. chemicals in your skin
   d. platelets in your blood

   ANS: A DIF: I OBJ: 6-2.1

8. Which of the following statements is true about the immune system?
   a. The immune system is localized in your body.
   b. The immune system is controlled by your brain.
   c. The immune system is an army of individual cells that fight pathogens.
   d. The immune system is controlled by your blood flow.

   ANS: C DIF: I OBJ: 6-2.2

9. All of the following cause a noninfectious disease EXCEPT
   a. a lack of vitamin C in the diet.
   b. a genetic disorder.
   c. some protists.
   d. most cancers.

   ANS: C DIF: I OBJ: 6-1.1

10. Cells that divide rapidly at an uncontrolled rate can cause
    a. multiple sclerosis.
    b. cancer.
    c. lupus.
    d. Graves' disease.

    ANS: B DIF: I OBJ: 6-3.2

11. Antigens attract
    a. memory B cells.
    b. killer T cells.
    c. pathogens.
    d. helper T cells.

    ANS: D DIF: I OBJ: 6-2.2

12. People with ____ have very few helper T cells.
    a. diabetes
    b. rheumatoid arthritis
    c. small pox
    d. AIDS

    ANS: D DIF: I OBJ: 6-3.3

13. A(n) ____ is NOT spread from one person to another.
    a. infectious disease
    b. noninfectious disease
    c. pathogen
    d. cold

    ANS: B DIF: I OBJ: 6-1.1

14. Infectious diseases are caused by agents called
    a. amino acids.
    b. hormones.
    c. pathogens.
    d. carcinogens.

    ANS: C DIF: I OBJ: 6-1.1

15. To avoid spreading pathogens, you should
    a. not drink from the same cup as your friend.
    b. cover your mouth completely while sneezing or coughing.
    c. refrigerate foods until you need them.
    d. All of the above

    ANS: D DIF: II OBJ: 6-1.2

16. After using a public restroom, a man leaves without washing his hands. How might others come into contact with a pathogen he could be carrying?
    a. the doorknob
    b. the faucet
    c. the towel
    d. the floor

    ANS: A DIF: II OBJ: 6-1.2

17. Which of the following is a way that pathogens might be transmitted at school through person-to-person contact?
    a. sharing a drink
    b. opening a locker
    c. talking to a nearby friend
    d. shaking hands in greeting

    ANS: D DIF: II OBJ: 6-1.2

18. After returning to your campsite from a hike in the woods, you should check yourself for ticks because ticks can carry and pass on
    a. hormones.
    b. pathogens.
    c. immunity.
    d. vaccines.

    ANS: B DIF: II OBJ: 6-1.2

19. It is important to wash all used cooking tools because
    a. the bacteria *Salmonella* could grow and produce toxins in food.
    b. you could get rabies.
    c. used cooking tools might pass on pathogens through the air.
    d. the dirty cooking tools might pass on a noninfectious disease.

    ANS: A DIF: II OBJ: 6-1.2

20. If you travel to a developing country, you may be advised to drink bottled water because the water supply might contain
    a. salts.
    b. minerals.
    c. pathogens.
    d. pesticides.

    ANS: C DIF: I OBJ: 6-1.2

21. Protection or resistance to a disease is called an
a. invalidation.
b. immunity.
c. antidote.
d. vaccine

ANS: B DIF: I OBJ: 6-1.3

22. A substance that helps your body develop immunity to a disease is a(n)
a. elixir.
b. antibiotic.
c. antidote.
d. vaccine.

ANS: D DIF: I OBJ: 6-1.3

23. Modern vaccines contain
a. pathogens.
b. antibiotics.
c. hormones.
d. vitamins and amino acids.

ANS: A DIF: I OBJ: 6-1.3

24. A substance that can kill bacteria or slow the growth of bacteria is a(n)
a. antidote.
b. antibiotic.
c. elixir.
d. vaccine.

ANS: B DIF: I OBJ: 6-1.3

25. Organisms that try to enter your eyes or mouth are usually destroyed by
a. hormones.
b. vaccines.
c. enzymes.
d. antibodies.

ANS: C DIF: I OBJ: 6-2.1

26. Pathogens that enter your nose are washed down the back of your throat by
a. hormones.
b. mucus.
c. enzymes.
d. antibodies.

ANS: B DIF: I OBJ: 6-2.1

27. The advantage of flushing pathogens into your stomach is that the pathogens are quickly
a. vomited out of the body.
b. removed from the body by white blood cells.
c. digested.
d. pasteurized.

ANS: C DIF: I OBJ: 6-2.1

28. Most pathogens that land on your skin cannot infect you because
a. they cannot find a live cell to infect.
b. they are quickly digested.
c. oils form a slippery surface that pathogens cannot cling to.
d. All of the above

ANS: A DIF: I OBJ: 6-2.1

29. As new skin cells grow from the inner layers, the outer layers of the skin
    a. become hard and calloused.
    b. become soft and supple
    c. become oily.
    d. flake off.

ANS: D   DIF: I   OBJ: 6-2.1

30. As old, dead skin is removed, it carries away
    a. moisture.
    b. pathogens.
    c. oil.
    d. oxygen.

ANS: B   DIF: I   OBJ: 6-2.1

31. Glands secrete ____ onto your skin that kill many pathogens.
    a. acidic water
    b. alkaline water
    c. oil
    d. T cells

ANS: C   DIF: I   OBJ: 6-2.1

32. Pathogens can enter the body when the skin is
    a. cut.
    b. burned.
    c. punctured.
    d. All of the above

ANS: D   DIF: I   OBJ: 6-2.2

33. ____ help to quickly seal an open wound so that no more pathogens can enter.
    a. T cells
    b. B cells
    c. Platelets
    d. Red blood cells

ANS: C   DIF: I   OBJ: 6-2.2

34. Increased blood flow to a wounded area brings cells that belong to the ____ system.
    a. integumentary
    b. nervous
    c. lymphatic
    d. immune

ANS: D   DIF: I   OBJ: 6-2.2

35. The immune system consists mainly of ____ types of cells
    a. two
    b. three
    c. four
    d. six

ANS: B   DIF: I   OBJ: 6-2.2

36. ____ eat any microorganisms or viruses floating around.
    a. Microphages
    b. Macrophages
    c. Micropathogens
    d. Macropathogens

ANS: B   DIF: I   OBJ: 6-2.2

37. If only a few microorganisms and viruses have entered the wound, the ____ can easily stop them.
a. T cells
b. B cells
c. antibiotics
d. macrophages

ANS: D DIF: I OBJ: 6-2.2

38. ____ play an important role in coordinating the immune system.
a. T cells
b. B cells
c. Antigens
d. Macrophages

ANS: A DIF: I OBJ: 6-2.2

39. Proteins that attach to specific pathogens are called
a. antibiotics.
b. B cells.
c. antibodies.
d. macrophages.

ANS: C DIF: I OBJ: 6-2.2

40. An antibody's shape is very specialized. It matches a ____ like a key fits a lock.
a. antigen
b. pathogen
c. macrophage
d. vaccine

ANS: B DIF: I OBJ: 6-2.2

41. To start the immune system response, a macrophage
a. engulfs a pathogen
b. sticks parts of the pathogen to its surface.
c. calls in killer T cells.
d. Both (a) and (b)

ANS: D DIF: I OBJ: 6-2.2

42. What activates helper T cells?
a. platelets helping to clot blood
b. macrophages wearing antigens
c. killer T cells attacking an infected cell
d. B cells making antibodies

ANS: B DIF: I OBJ: 6-2.2

43. Who notifies the killer T cells that pathogens are present?
a. helper T cells
b. B cells
c. macrophages
d. antigens

ANS: A DIF: I OBJ: 6-2.2

44. How are helper T cells able to move quickly and efficiently?
   a. Helper T cells travel only in a major artery.
   b. Helper T cells quickly divide.
   c. Helper T cells send electrical messages along nerves.
   d. all of the above

ANS: B DIF: I OBJ: 6-2.2

45. Killer T cells kill
   a. antigens.
   b. pathogens.
   c. cells infected with pathogens.
   d. All of the above

ANS: C DIF: I OBJ: 6-2.2

46. Cells infected with pathogens
   a. run to T cells for help.
   b. place antigens on their surface.
   c. die instantly.
   d. are engulfed by macrophages.

ANS: B DIF: I OBJ: 6-2.2

47. Who notifies the B cells that pathogens are present?
   a. helper T cells
   b. killer T cells
   c. macrophages
   d. antigens

ANS: A DIF: I OBJ: 6-2.2

48. ____ make antibodies that cling to many of the pathogens.
   a. Macrophages
   b. Killer T cells
   c. Helper T cells
   d. B cells

ANS: D DIF: I OBJ: 6-2.2

49. Once a pathogen has a(n) ____ attached to it, all kinds of immune cells and proteins swarm to attack the pathogen.
   a. killer T cell
   b. antigen
   c. antibody
   d. B cell

ANS: C DIF: I OBJ: 6-2.2

50. Special ____ attach themselves to the antibodies stuck on pathogens. They then punch holes in the pathogens, destroying them.
   a. helper T cells
   b. killer T cells
   c. proteins
   d. B cells

ANS: C DIF: I OBJ: 6-2.2

51. ____ engulf pathogens tagged with antibodies.
   a. B cells
   b. Macrophages
   c. Killer T cells
   d. Antigens

ANS: B DIF: I OBJ: 6-2.2

52. When macrophages activate helper T cells, they also send a chemical signal that tells your brain to

a. increase the temperature.
b. decrease the temperature.
c. remember how to make an antibody.
d. make more antibodies.

ANS: A DIF: I OBJ: 6-2.3

53. If your immune system is so efficient, why do you get sick?

a. B cells must have had previous contact with a pathogen before they can make the correct antibodies.
b. When encountering a new pathogen, specialized B cells make antibodies that are effective against that particular invader.
c. B cells take about 2 weeks to develop a new antibody, which is too long to prevent an infection.
d. all of the above

ANS: D DIF: I OBJ: 6-2.2

54. Which of the following can NOT cause an allergic reaction by the immune system?

a. certain foods
b. pollen
c. cancerous cells
d. dander

ANS: C DIF: I OBJ: 6-3.1

55. Scientists think allergies might be useful because

a. the mucus draining from your nose carries away irritants.
b. it gives your immune system practice for a real pathogen.
c. immune cells need the experience.
d. None of the above

ANS: A DIF: I OBJ: 6-3.1

56. Autoimmune diseases occur when the immune system

a. detects pathogens.
b. detects antigens.
c. cannot distinguish between pathogens and body cells.
d. can no longer produce macrophages.

ANS: C DIF: I OBJ: 6-3.1

57. Healthy cells divide

a. when they are fertilized.
b. at a carefully regulated rate.
c. randomly.
d. uncontrollably.

ANS: B DIF: I OBJ: 6-3.2

58. Cancer is deadly because killer T cells
    a. are incapacitated.
    b. cannot destroy the unregulated cells fast enough.
    c. cannot destroy unregulated cells.
    d. are used as factories to make the virus.

    ANS: B   DIF: I   OBJ: 6-3.2

59. The progress of cancer can be slowed by
    a. killer T cells.
    b. radiation.
    c. certain drugs.
    d. All of the above

    ANS: D   DIF: I   OBJ: 6-3.2

60. The acronym HIV stands for
    a. human immunodeficiency virus.
    b. homo infectory virus.
    c. homo immunodeficiency virus.
    d. human infectory virus.

    ANS: A   DIF: I   OBJ: 6-3.3

61. The acronym AIDS stands for
    a. auto immune deficiency syndrome.
    b. acquired immune deficiency syndrome.
    c. active immune deficiency symptoms.
    d. auto infective deficiency symptoms.

    ANS: B   DIF: I   OBJ: 6-3.3

62. HIV primarily infects
    a. T cells.
    b. B cells.
    c. macrophages.
    d. unregulated cells.

    ANS: A   DIF: I   OBJ: 6-3.3

63. HIV destroys helper T cells by
    a. puncturing the T cells.
    b. surrounding the T cells and utterly destroying them.
    c. using the T cells as factories to produce more viruses.
    d. being absorbed by the T cells and then eating them from the inside out.

    ANS: C   DIF: I   OBJ: 6-3.3

64. The immune system of a person with AIDS cannot attack any pathogen because the
    a. helper T cells cannot activate other cells.
    b. killer T cells are incapacitated.
    c. macrophages are too slow to engulf them.
    d. B cells cannot produce the right antibodies.

    ANS: A   DIF: I   OBJ: 6-3.3

COMPLETION

1. Diseases caused by pathogens are ____________________. (infectious or noninfectious)

   ANS: infectious DIF: I OBJ: 6-1.1

2. Antibiotics can be used to kill some ____________________. (bacteria or viruses)

   ANS: bacteria DIF: I OBJ: 6-1.3

3. Macrophages attract ____________________. (helper T cells or killer T cells)

   ANS: helper T cells DIF: I OBJ: 6-2.2

4. Certain B cells make ____________________. (antigens or antibodies)

   ANS: antibodies DIF: I OBJ: 6-2.2

5. An immune system overreaction to a harmless substance is a(n) ____________________. (allergy or vaccine)

   ANS: allergy DIF: I OBJ: 6-3.1

6. ____________________ attacks helper T cells. (HIV or Cancer)

   ANS: HIV DIF: I OBJ: 6-3.3

7. ________________________ are caused by pathogens, such as bacteria and viruses. (Infectious diseases or Noninfectious diseases)

   ANS: Infectious diseases DIF: I OBJ: 6-1.1

8. ____________________ are responsible for telling the brain to increase the body temperature during infection. (Antibodies or Macrophages)

   ANS: Macrophages DIF: I OBJ: 6-2.3

9. An ____________________ is a reaction that occurs when the immune system overreacts to antigens that are not dangerous to the body. (immunity or allergy)

   ANS: allergy DIF: I OBJ: 6-3.1

10. ____________________ make antibodies. (T cells or B cells)

    ANS: B cells DIF: I OBJ: 6-2.2

11. Doctors usually treat bacterial infections with a(n) ____________________. (antibiotic or pathogen)

ANS: antibiotic DIF: I OBJ: 6-1.3

12. The use of heat to kill bacteria in food and beverages is called ____________________. (immunity or pasteurization)

ANS: pasteurization DIF: I OBJ: 6-1.3

13. An ____________________ disease can be passed from one living thing to another.

ANS: infectious DIF: I OBJ: 6-1.1

14. ____________________ are often referred to as germs.

ANS: Pathogens DIF: I OBJ: 6-1.1

15. ____________________ are not affected by antibiotics.

ANS: Viruses DIF: I OBJ: 6-1.3

16. The ____________________ system is the body system that fights pathogens.

ANS: immune DIF: I OBJ: 6-2.2

17. ____________________ are pieces of pathogens that are displayed on macrophages.

ANS: Antigens DIF: I OBJ: 6-2.2

18. ____ "remember" how to make an antibody for a particular pathogen.

ANS: Memory B cells DIF: I OBJ: 6-2.2

19. An ____________________ is when the immune system overreacts to antigens that are NOT dangerous to the body.

ANS: allergy DIF: I OBJ: 6-2.2

20. An ____ is a disease in which the immune system attacks the body's own cells

ANS: autoimmune disease DIF: I OBJ: 6-3.1

21. ____________________ occurs when a cell does NOT respond to the body's regulation and begins dividing at an uncontrolled rate.

ANS: Cancer DIF: I OBJ: 6-3.2

22. In rheumatoid arthritis, immune-system cells cause joint-tissue swelling, which can lead to joint deformities. Rheumatoid arthritis is a(n) ____________________.

ANS: autoimmune disease DIF: I OBJ: 6-1.1

SHORT ANSWER

1. How might an insect transmit a pathogen from water to a human?

ANS:
Sample answer: The insect lands on the water and ingests the pathogen. The insect carries the pathogen to a human and injects the pathogen into the human host.

DIF: II OBJ: 6-1.2

2. How is an infectious disease different from a noninfectious disease?

ANS:
An infectious disease can be transmitted to other organisms. A noninfectious disease cannot be transmitted to another organism.

DIF: I OBJ: 6-1.1

3. List five ways that you might come into contact with a pathogen.

ANS:
Answers can vary, but may include being sneezed upon by an infected person, touching an infected person, having sexual contact with an infected person, eating infected food, or breathing in the pathogen.

DIF: I OBJ: 6-1.2

4. How does a vaccine work?

ANS:
A vaccine is enough like a pathogen to help a body develop a defense against that pathogen.

DIF: I OBJ: 6-1.3

5. Why might the risk of infectious disease be high in a community that has no water-treatment facility?

ANS:
Without water treatment, a community's water supply can become easily infected with pathogens. These pathogens can be ingested or can contaminate objects washed in the water.

DIF: II OBJ: 6-1.2

6. List three ways your body defends itself against pathogens.

ANS:
Sample answer: Skin keeps most pathogens out. Chemical defenses are in your eyes, stomach, and mouth. If you have a cut in your skin, blood platelets help to close the wound so more microorganisms cannot enter. Any microorganisms that do enter encounter immune system cells.

DIF: I OBJ: 6-2.1

7. Name three different cells in the immune system, and describe how they respond to pathogens.

ANS:
Sample answer: Macrophages engulf pathogens and stick antigens on their outer membranes. Helper T cells activate killer T cells and B cells. Killer T cells kill any body cell infected with pathogens.

DIF: I OBJ: 6-2.2

8. Can you make antibodies for diseases you have never come in contact with? Why or why not?

ANS:
No; B cells can't make antibodies for a disease until they have encountered that disease at least once before.

DIF: I OBJ: 6-2.2

9. If you had chickenpox at age 7, what would be your chances of getting chickenpox again if your memory B cells lived only 2 months?

ANS:
You could get chickenpox every other month, because there would be no memory B cells to "remember" the pathogen.

DIF: II OBJ: 6-2.2

10. What is the difference between allergies and autoimmune diseases?

ANS:
In allergies, the immune system attacks harmless antigens. In autoimmune diseases, the immune system attacks healthy body cells.

DIF: I OBJ: 6-3.1

11. Why is it important for immune-system cells to be able to recognize all of the body's own cells?

ANS:
Immune system cells must be able to recognize the body's own cells so they can determine when the body has been invaded.

DIF: I OBJ: 6-3.3

12. What characterized a cancerous cell?

ANS:
A cell becomes cancerous when it starts dividing at an uncontrolled rate.

DIF: I OBJ: 6-3.2

13. Over time, people with AIDS become very sick and are unable to fight off infection. Use the information in the graph below to explain why this occurs.

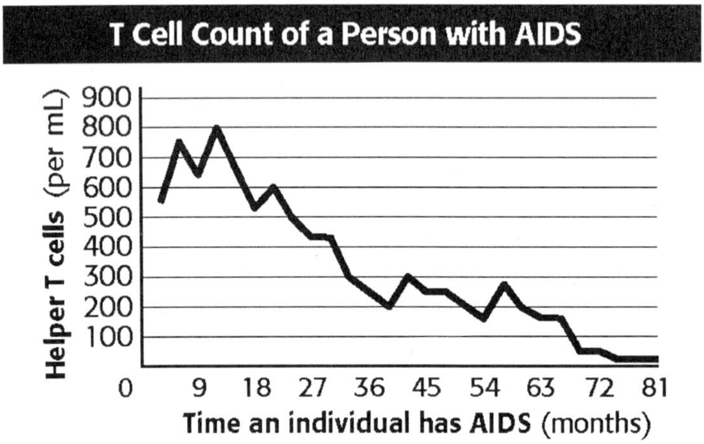

ANS:
People with AIDS become less able to fight infections because their helper T cell count decreases over time.

DIF: II OBJ: 6-3.3

14. Explain how macrophages start an immune response.

ANS:
When a macrophage engulfs a pathogen, it places pieces of the pathogen called antigens on its outer membrane. The antigens attract helper T cells.

DIF: I OBJ: 6-2.2

15. Describe the role of helper T cells in responding to an infection.

ANS:
Helper T cells activate B cells and killer T cells.

DIF: I OBJ: 6-2.2

16. Use the following terms to create a concept map: *macrophages, helper T cells, B cells, antibodies, antigens, killer T cells, memory B cells.*

ANS:

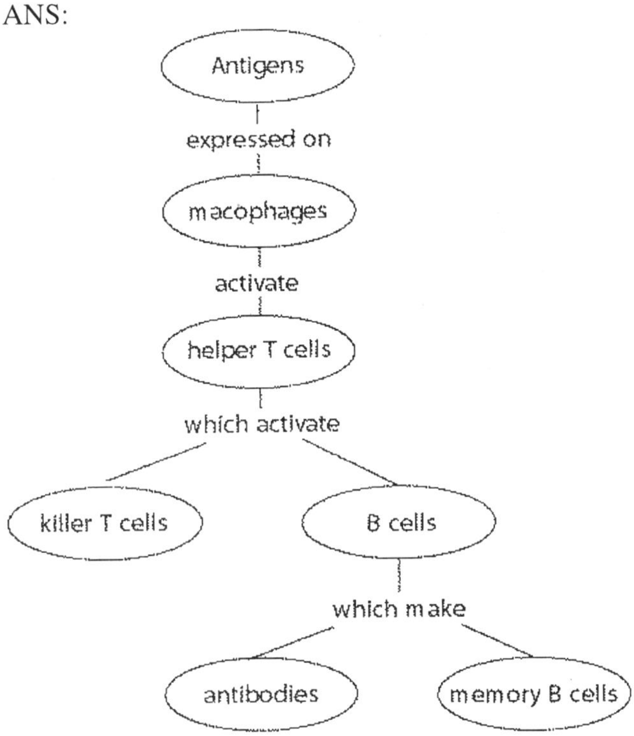

DIF: II OBJ: 6-2.2

17. Why does the disappearance of helper T cells in AIDS patients damage the immune system?

ANS:
Helper T cells are needed to activate B cells and killer T cells. Without B cells, the immune system cannot produce antibodies. Without killer T cells, the immune system cannot destroy infected cells.

DIF: II OBJ: 6-3.3

18. Many people take fever-reducing drugs as soon as their temperature exceeds 37°C.
    a. Why might it not be a good idea to immediately reduce a fever with drugs?
    b. What are the benefits of taking fever-reducing drugs?

ANS:

a. Fevers can help T cells and B cells multiply faster. Fevers can also slow the growth of some pathogens. High fevers can be dangerous, though.
b. If a fever exceeds 40.6°C, fever-reducing drugs can help bring the body's temperature down to a safer temperature.

DIF: II OBJ: 6-2.3

19. The risk of dying from a whooping cough vaccine is about one in 1 million. In contrast, the risk of dying from whooping cough itself is about one in 500. Discuss the pros and cons of this vaccination.

ANS:
The vaccine gives you an immunity to whooping cough, which is a dangerous disease. However, you may never come into contact with whooping cough and, therefore, may not need the vaccine. If you receive the vaccine anyway, you will have a small risk of dying from the vaccine. If no one is vaccinated, however, the risk of contracting the disease rises.

DIF: II OBJ: 6-1.3

20. Suppose you have 50,000 flu viruses on your fingers and you rub your eyes. Only 20,000 viruses make it into your eyes, 10,000 dissolve in chemicals, and 10,000 are washed down into your nose. Of those, you sneeze out 2,000. How many viruses are left to wash down the back of your throat and start an infection?

ANS:
Only 8,000 viruses remain to start an infection.

DIF: II OBJ: 6-1.3

The graph below compares the concentration of antibodies in the blood the first time you are exposed to a pathogen with the concentration of antibodies the next time you are exposed to the pathogen.

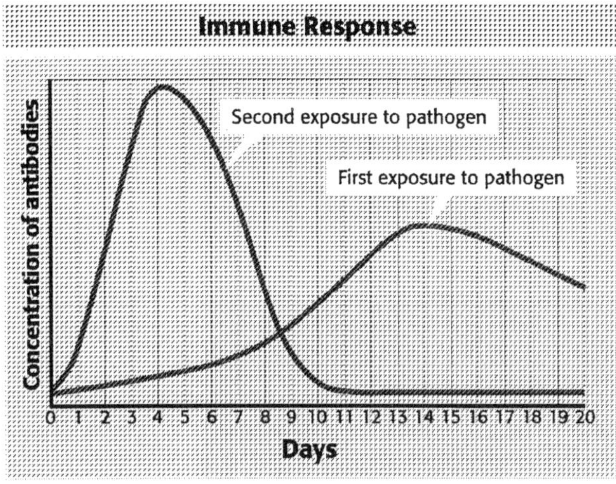

21. Are there more antibodies present during the first week of the first exposure or the first week of the second exposure? Why do you think this is so?

ANS:
There are more antibodies present in the second exposure because memory B cells recognize the pathogen and immediately start the production of antibody-producing B cells.

DIF: II OBJ: 6-2.2

22. What is the difference in recovery time between the first exposure and second exposure? Why?

ANS:
Recovery time is much shorter for the second exposure because your body can produce more antibodies in less time than during the first exposure.

DIF: II OBJ: 6-2.2

23. Why are memory B cells important?

ANS:
Memory B cells recognize pathogens that have invaded the body before, and they make antibodies to fight off these pathogens. When infected, individuals may never feel sick because memory B cells immediately recognize the pathogen and destroy it.

DIF: I OBJ: 6-2.2

24. Explain how vaccines work.

ANS:
Vaccines contain pathogens that are dead or specially treated so that they do not make you sick. Vaccines allow your body to develop defenses against the disease so if the actual pathogen enters the body, the pathogen is less harmful or not harmful at all.

DIF: I OBJ: 6-1.3

25. Name four ways a waiter could contract an infectious disease at work.

ANS:
Sample answer: The waiter could contract an infectious disease by touching another person directly, by touching an object such as money that an infected person previously touched, by simply breathing the same air as an infected person, or by drinking from the same glass an infected person has used.

DIF: II OBJ: 6-1.2

26. Use the following terms to complete the concept map below: *B cells, pathogens, antigens, T cells, killer T cells, macrophages.*

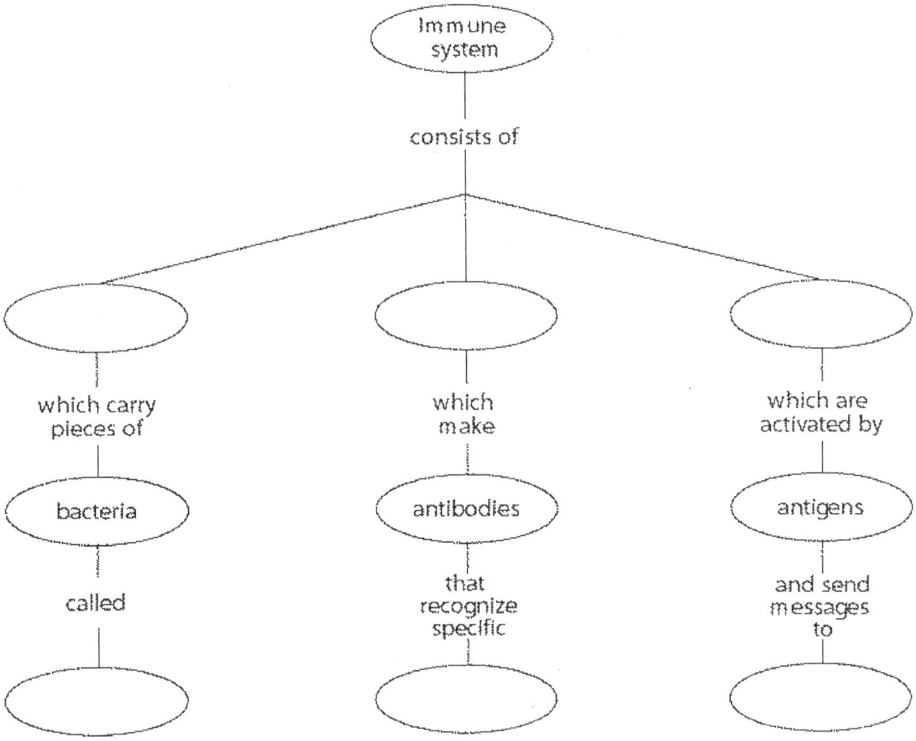

ANS:

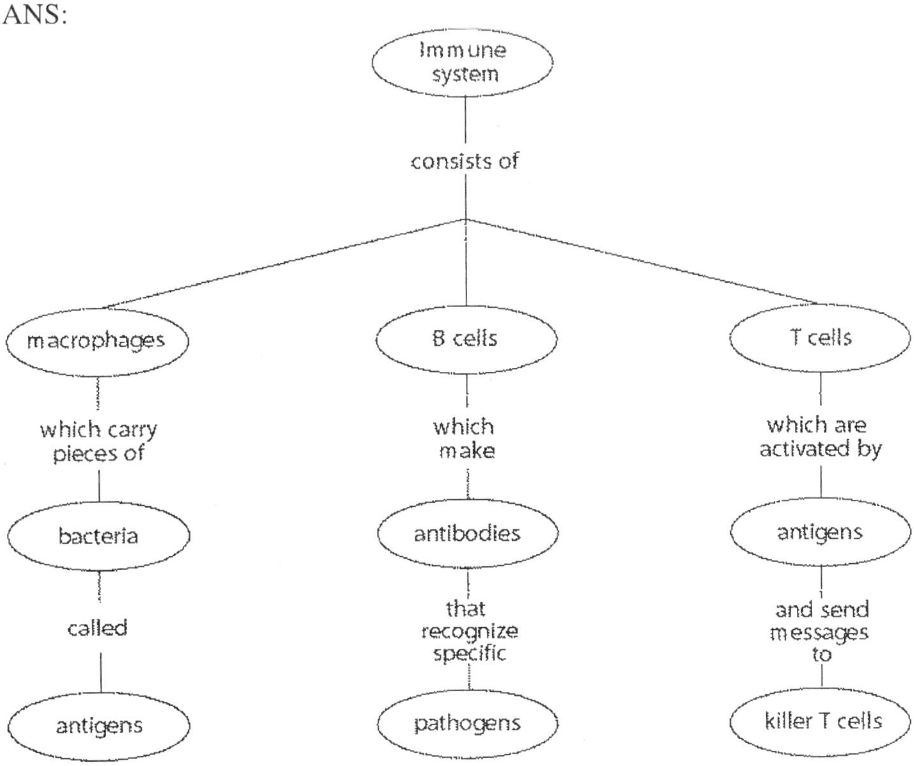

DIF: II OBJ: 6-2.2

Examine the diagram below, and answer the questions that follow.

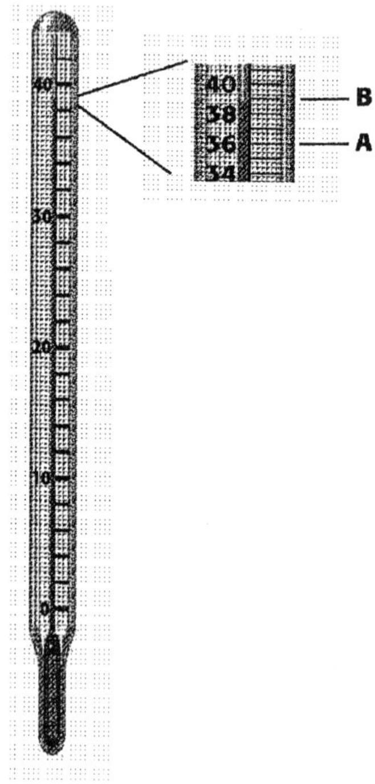

27. What does A represent?

ANS:
A represents normal body temperature.

DIF: II OBJ: 6-2.3

28. What does B represent?

ANS:
B represents the best temperature for T cells and B cells to reproduce.

DIF: II OBJ: 6-2.3

29. Why does your body temperature change when a pathogen invades?

ANS:
Your body temperature increases when a pathogen invades so that T cells and B cells multiply faster. More T cells and B cells help the immune system fight off the pathogen.

DIF: II OBJ: 6-2.3

30. Raw, unpasteurized milk contains enzymes that die if they are heated to a temperature of 54°C. The process of pasteurization heats milk to a range of 63–77°C. What are some advantages and disadvantages of pasteurization?

ANS:
The *advantage* of pasteurization is that it kills pathogens. The *disadvantage* is that it destroys enzymes.

DIF: III OBJ: 6-1.3

31. Your friend's doctor prescribes an antibiotic that your friend must take twice a day for 1 week. When your friend feels better after only 2 days, she wants to stop taking the antibiotic. What would you say to convince your friend to continue taking the antibiotic?

ANS:
Sample answer: I would tell my friend that she is feeling better because the antibiotic is slowing the growth of the pathogen. However, to kill all the pathogens she needs to finish the antibiotic. If she stops now, the remaining pathogens will reproduce and make her sick again.

DIF: III OBJ: 6-1.2

32. Describe the changes that would occur to the immune system response in a person who has no B cells.

ANS:
The immune system response would be greatly reduced in a person who has no B cells. B cells produce antibodies which tag pathogens for destruction. Macrophages and special proteins attack pathogens that are marked by these antibodies. Without B cells, the macrophages and proteins would not participate in this stage of the immune system response.

DIF: II OBJ: 6-2.2

33. Why is it important to thoroughly clean a cut?

ANS:
Cleaning a cut can get rid of some of the pathogens that could cause infection.

DIF: II OBJ: 6-1.3

34. Suppose you are fortunate enough to visit Peru. On your first day in Peru, you eat all kinds of new fruits such as granadillas and guanabanas. The next morning, you feel sick and have diarrhea. How might eating the fruit have caused you to come into contact with a pathogen that made you sick?

ANS:
Sample answers: The fruit was rinsed in tap water which contained pathogens. The fruit was handled by someone carrying the pathogen.

DIF: III OBJ: 6-1.2

35. People often take "boosters" of some vaccines at regular intervals. Why do you think a booster is helpful?

ANS:
Taking a booster will increase the number of memory B cells that produce the antibody for the disease.

DIF: III OBJ: 6-2.2

36. If your immune system could NOT form memory B cells, would vaccines protect you from pathogens? Explain your answer.

ANS:
No, vaccines cause the body to make memory B cells against specific pathogens. Without memory B cells, vaccines would provide no long-term protection against the pathogens.

DIF: III OBJ: 6-2.2

37. When a person receives an organ from another person, his or her immune system often begins to attack the new organ. Why do you think this happens?

ANS:
The immune system recognizes the organ as foreign—not part of the body—and begins to attack it as if it were a pathogen or an unregulated cell.

DIF: III OBJ: 6-2.2

38. Would HIV's effects on the immune system be as damaging if it invaded only killer T cells? Explain your answer.

ANS:
The effects would not be as great since the body would still have macrophages to attack pathogens, and helper T cells would be free to activate B cells, resulting in antibody production.

DIF: II OBJ: 6-3.3

39. How might an insect transmit a pathogen from water to a human?

ANS:
Sample answer: The insect lands on the water and ingests the pathogen. The insect carries the pathogen to a human and injects the pathogen into the human host.

DIF: II OBJ: 6-1.2

40. Compost piles often reach temperatures of at least 40°C. Compare this temperature with the temperature at which your body best fights off infection.

ANS:
A temperature of 40°C is higher than the temperature at which my body best fights off infection, which is 39°C.

DIF: III OBJ: 6-2.3

41. People who receive severe burns are at an increased risk of infection. Considering what you know about the body's defenses, explain how this is possible.

ANS:
The first line of defense is weakened because skin damaged by burns does not provide the same protection against pathogens as healthy skin.

DIF: II OBJ: 6-2.1

42. Why is it better for a pathogen not to kill its host?

ANS:
It is better for a pathogen not to kill its host because the death of the host hinders or stops the transfer of the pathogen to a new host.

DIF: III OBJ: 6-2.3

MULTIPLE CHOICE

1. Which of the following is NOT a function of water in the body?
   a. transport substances
   b. regulate temperature
   c. provide Calories
   d. provide lubrication

   ANS: C DIF: I OBJ: 7-1.1

2. Side effects of over-the-counter and prescription medicines may include
   a. nausea.
   b. headaches.
   c. drowsiness.
   d. All of the above

   ANS: D DIF: I OBJ: 7-2.1

3. Which of the following nutrients does NOT provide energy for the body?
   a. carbohydrates
   b. vitamins
   c. fats
   d. proteins

   ANS: B DIF: I OBJ: 7-1.1

4. What are the effects of nicotine on the body?
   a. liver damage and decrease in physical endurance
   b. loss of appetite and decrease in physical endurance
   c. loss of appetite and brain damage
   d. liver damage and loss of appetite

   ANS: B DIF: I OBJ: 7-2.2

5. Which of the following statements about drugs is true?
   a. All drugs are illegal.
   b. Smoking just one or two cigarettes is safe for anyone.
   c. Alcohol is not a drug.
   d. Withdrawal symptoms may be painful.

   ANS: D DIF: I OBJ: 7-2.2

6. Aerobic exercise does NOT
   a. burn calories.
   b. increase the heart rate.
   c. strengthen the heart, lungs, and bones.
   d. make you weak.

   ANS: D DIF: I OBJ: 7-3.2

7. When talking to a 911 operator, you should NOT
   a. describe the accident.
   b. ask what to do.
   c. hang up before the dispatcher hangs up.
   d. speak slowly and clearly.

   ANS: C DIF: I OBJ: 7-3.4

8. Which of the following provides energy for the body in the form of Calories?
   a. vitamins
   b. water
   c. proteins
   d. minerals

   ANS: C DIF: I OBJ: 7-1.1

9. Fats do NOT
   a. help skin stay healthy.
   b. protect internal organs.
   c. regulate hormone levels.
   d. insulate the body.

   ANS: C DIF: I OBJ: 7-1.1

10. ____ is NOT a nutrient.
   a. Fiber
   b. A vitamin
   c. Water
   d. A mineral

   ANS: A DIF: I OBJ: 7-1.1

11. Your body is mostly composed of
   a. minerals.
   b. water.
   c. vitamins.
   d. muscle.

   ANS: B DIF: I OBJ: 7-1.1

12. ____ are drugs that relax muscles and help people sleep.
   a. Stimulants
   b. Antihistamines
   c. Analgesics
   d. Sedatives

   ANS: D DIF: I OBJ: 7-2.1

13. Tobacco is NOT
   a. addictive.
   b. less hazardous if it is the smokeless variety.
   c. linked to decreased physical endurance.
   d. linked to heart disease.

   ANS: B DIF: I OBJ: 7-2.2

14. Which of the following is NOT true about drugs?
    a. Drugs can only enter the body if they are swallowed.
    b. Drugs can cause emotional, physical, and behavioral changes.
    c. Alcohol is a type of drug.
    d. Drugs can help people recover from an illness.

    ANS: A DIF: I OBJ: 7-2.1

15. In order to stay healthy, you need to take in more than 40 different
    a. nutrients.
    b. carbohydrates.
    c. proteins.
    d. minerals.

    ANS: A DIF: I OBJ: 7-1.1

16. Which of the following is NOT a nutrient?
    a. carbohydrate
    b. DNA
    c. vitamin
    d. mineral

    ANS: B DIF: I OBJ: 7-1.1

17. The energy in nutrients is measured in units called
    a. calories.
    b. Calories.
    c. Joules.
    d. grams.

    ANS: B DIF: I OBJ: 7-1.1

18. A ____ is a chemical composed of one or more simple sugars.
    a. protein
    b. fat
    c. carbohydrate
    d. vitamin

    ANS: C DIF: I OBJ: 7-1.1

19. Which of the following is your body's main source of energy?
    a. water
    b. fats
    c. proteins
    d. carbohydrates

    ANS: D DIF: I OBJ: 7-1.1

20. A major source of carbohydrates is
    a. plants.
    b. fish.
    c. milk.
    d. nuts.

    ANS: A DIF: I OBJ: 7-1.1

21. ____ are sugars that are easily digested and give your body quick energy.
    a. Simple carbohydrates
    b. Complex carbohydrates
    c. Proteins
    d. Fats

    ANS: A DIF: I OBJ: 7-1.1

22. ____ are made of many sugar molecules that are linked together.

a. Proteins
b. Complex carbohydrates
c. Simple carbohydrates
d. Fats

ANS: B DIF: I OBJ: 7-1.1

23. ____ are nutrients used to build and repair body parts.

a. Simple carbohydrates
b. Fats
c. Proteins
d. Complex carbohydrates

ANS: C DIF: I OBJ: 7-1.1

24. Your body makes the proteins it needs from building blocks called

a. nutrients.
b. amino acids.
c. DNA.
d. carbohydrates.

ANS: B DIF: I OBJ: 7-1.1

25. ____ are energy-storage nutrients that help the body store some vitamins.

a. Fats
b. Amino acids
c. Proteins
d. Carbohydrates

ANS: A DIF: I OBJ: 7-1.1

26. ____ are found in meats, dairy products, coconut oil, and palm oil.

a. Complex carbohydrates
b. Simple carbohydrates
c. Saturated fats
d. Unsaturated fats

ANS: C DIF: I OBJ: 7-1.1

27. ____ are found in vegetable oils and fish.

a. Saturated fats
b. Unsaturated fats
c. Simple carbohydrates
d. Complex carbohydrates

ANS: B DIF: I OBJ: 7-1.1

28. ____ are known to raise blood cholesterol levels.

a. Simple carbohydrates
b. Complex carbohydrates
c. Saturated fats
d. Unsaturated fats

ANS: C DIF: I OBJ: 7-1.1

29. ____ are elements that are essential for good health.

a. Proteins
b. Vitamins
c. Minerals
d. Nucleic acids

ANS: C DIF: I OBJ: 7-1.1

30. Which of the following minerals are necessary for strong bones and teeth?
a. potassium and sodium
b. fluorine and iodine
c. calcium and magnesium
d. calcium and potassium

ANS: C DIF: I OBJ: 7-1.1

31. Which of the following minerals is needed to regulate your heartbeat and produce muscle movement?
a. potassium
b. sodium
c. magnesium
d. calcium

ANS: A DIF: I OBJ: 7-1.1

32. Which of the following minerals is necessary for red blood cell production?
a. potassium
b. sodium
c. iron
d. zinc

ANS: C DIF: I OBJ: 7-1.1

33. ____, like aspirin, are pain-relieving drugs.
a. Antibiotics
b. Analgesics
c. Antihistamines
d. Sedatives

ANS: B DIF: I OBJ: 7-2.1

34. ____ fight bacterial infections.
a. Antibiotics
b. Analgesics
c. Antihistamines
d. Sedatives

ANS: A DIF: I OBJ: 7-2.1

35. ____ help control symptoms of allergies, asthma, and colds.
a. Antibiotics
b. Analgesics
c. Antihistamines
d. Sedatives

ANS: C DIF: I OBJ: 7-2.1

36. Regular use of some drugs can cause the body to develop
a. obesity.
b. anorexia nervosa.
c. tolerance.
d. malnutrition.

ANS: C DIF: I OBJ: 7-2.1

37. ____ are drugs made from opium.
a. Hallucinogens
b. Narcotics
c. Barbiturates
d. Amphetamines

ANS: B DIF: I OBJ: 7-2.2

38. ____ refers to methods of preserving and protecting your health.

a. Drug use
b. Alcoholism
c. Hygiene
d. none of the above

ANS: C DIF: I OBJ: 7-3.1

39. The best way to prevent the spread of disease and infection is to

a. stay indoors.
b. take lots of antibiotics.
c. use tissue.
d. wash your hands.

ANS: D DIF: I OBJ: 7-3.1

40. Using ____ can help prevent sunburn, wrinkles, and skin cancer.

a. water
b. sunscreen
c. lotion
d. prescription drugs

ANS: B DIF: I OBJ: 7-3.1

41. Which of the following should you do for good oral hygiene?

a. floss at least once a day
b. get regular dental exams
c. eat a healthy diet
d. all of the above

ANS: D DIF: I OBJ: 7-3.1

42. Why is good posture important to health?

a. Good posture makes it difficult to breathe.
b. Good posture strains your muscles and ligaments.
c. Good posture helps your body look and feel good.
d. all of the above

ANS: C DIF: I OBJ: 7-3.1

43. Regular aerobic exercise at least ____ times a week is critical to good health.

a. two
b. three
c. five
d. seven

ANS: B DIF: I OBJ: 7-3.2

44. Aerobic exercise is vigorous, sustained exercise of the whole body for ____ or more.

a. 10 minutes
b. 15 minutes
c. 20 minutes
d. 1 hour

ANS: C DIF: I OBJ: 7-3.2

45. Aerobic exercise causes more

a. oxygen to be distributed throughout the body.
b. fatigue.
c. stress.
d. None of the above

ANS: A DIF: I OBJ: 7-3.2

46. Teenagers need ____ children.
   a. the same amount of sleep as
   b. less sleep than
   c. more sleep than
   d. hardly any sleep when compared with

   ANS: C DIF: I OBJ: 7-3.2

47. Teenagers need about ____ hours of sleep each night.
   a. 4
   b. 6
   c. 8
   d. 9.5

   ANS: D DIF: I OBJ: 7-3.2

48. Which of the following is a good idea for handling stress?
   a. Work harder.
   b. Exercise regularly, and get enough sleep.
   c. Drink caffeinated beverages.
   d. Eat chocolate.

   ANS: B DIF: I OBJ: 7-3.3

49. Never touch ____ while touching water.
   a. books
   b. nonslip mats
   c. electrical switches or appliances
   d. cold metals

   ANS: C DIF: I OBJ: 7-3.4

50. To be safe in case of a household fire, you should have ____ on every floor.
   a. the telephone number 911
   b. smoke detectors
   c. carbon monoxide detectors
   d. motion detectors

   ANS: B DIF: I OBJ: 7-3.4

51. To prevent a household fire, you should NOT
   a. plug too many appliances into one outlet.
   b. leave the water running.
   c. learn first aid.
   d. install smoke detectors.

   ANS: A DIF: I OBJ: 7-3.4

52. Why should you NOT walk away from a gas stove while heating oil in a skillet?
   a. Oil must be stirred constantly.
   b. The flame from the stove's burner could ignite the oil.
   c. The oil might not heat fast enough.
   d. Your recipe will be ruined.

   ANS: B DIF: I OBJ: 7-3.4

53. Which of the following is a safety hazard in the kitchen?
  a. allowing pot handles to extend over the edge of the stove
  b. cleaning up grease drippings immediately
  c. using a stool to reach high shelves
  d. all of the above

ANS: A DIF: I OBJ: 7-3.4

54. In an emergency, you should dial ____ on the telephone.
  a. 0
  b. 1411
  c. SOS
  d. 911

ANS: D DIF: I OBJ: 7-3.4

55. Which statement best reflects the safest thing you can do while in a car?
  a. Wear a helmet.
  b. Keep water with you.
  c. Wear a seat belt.
  d. Keep a cell phone with you.

ANS: C DIF: I OBJ: 7-3.4

56. If you ride your bicycle to school, you should ride ____ the flow of traffic.
  a. against
  b. in the middle of
  c. with
  d. opposite

ANS: C DIF: I OBJ: 7-3.4

57. Which of the following is a good outdoor safety habit?
  a. hiking alone
  b. learning how to swim
  c. camping alone
  d. drinking water from a stream

ANS: B DIF: I OBJ: 7-3.4

58. If you know CPR, will you need to call for help in an emergency?
  a. No, because CPR is all I would need to save the person's life.
  b. No, because CPR is the only training that the medics receive.
  c. Yes, because I might not do the CPR procedure correctly.
  d. Yes, because the person may need further medical attention.

ANS: D DIF: I OBJ: 7-3.4

59. Most teenage girls need about ____ Calories a day.
  a. 800
  b. 1,200
  c. 1,800
  d. 2,200

ANS: D DIF: I OBJ: 7-1.2

60. Most teenage boys need about ____ Calories a day.
  a. 900
  b. 1,800
  c. 2,800
  d. 3,200

ANS: C DIF: I OBJ: 7-1.2

61. Which of the following is one serving of rice, pasta, or cooked cereal?
    a. 1 oz
    b. 1/2 cup
    c. 3/4 cup
    d. 1 cup

    ANS: B   DIF: I   OBJ: 7-1.2

62. Which of the following is one serving of cooked dried beans?
    a. 1 oz
    b. 1/2 cup
    c. 3/4 cup
    d. 1 cup

    ANS: B   DIF: I   OBJ: 7-1.2

63. Which of the following is one serving of fruits?
    a. 1 medium apple, orange, or banana
    b. 1/2 cup of chopped fruit
    c. 3/4 cup of fruit juice
    d. all of the above

    ANS: D   DIF: I   OBJ: 7-1.2

64. Which of the following is one serving of chopped vegetables?
    a. 1/4 cup
    b. 1/2 cup
    c. 3/4 cup
    d. 1 cup

    ANS: B   DIF: I   OBJ: 7-1.2

Examine the nutrition information label and answer the questions that follow.

%Nutrition Facts	
Serving Size: 3/4 cup (30 g)	
Servings Per Container: about 12	
Amount Per Serving	
Calories	120
Calories from Fat	10
	% Daily Value
Total Fat 1.5 g	2%
Saturated Fat 0g	0%
Cholesterol 0mg	0%
Sodium 175mg	7%
Total Carbohydrate 24g	8%
Dietary Fiber 3g	12%
Sugars 7g	
Protein 3g	
Vitamin A	0%
Vitamin C	0%
Calcium	2%
Iron	6%

Ingredients: Organic Whole Oat Flour, Organic Corn Meal, Organic Evaporated Cane Juice, Organic Brown Rice Flour, Organic Barley Malt Extract, Sea Salt, Natural Blueberry Flavor.

65. Suppose you ate 1.5 cups of this food as an afternoon snack. How many Calories did you eat?
a. 120
b. 130
c. 240
d. 260

ANS: C DIF: II OBJ: 7-1.3

66. Which ingredient was used in the greatest amount?
a. Organic Whole Oat Flour
b. Organic Corn Meal
c. Sea Salt
d. Natural Blueberry Flavor

ANS: A DIF: II OBJ: 7-1.3

COMPLETION

1. ____________________ are linked to high blood cholesterol levels. (Saturated fats or Unsaturated fats)

ANS: Saturated fats DIF: I OBJ: 7-1.1

2. Physical dependence on a drug is called ____________________. (addiction or tolerance)

ANS: addiction DIF: I OBJ: 7-2.2

3. The ____________________ divides foods into six groups and gives a recommended number of servings for each. (nutrition information label or food pyramid)

ANS: food pyramid DIF: I OBJ: 7-1.2

4. One of the characteristics of ____________________ is binge eating. (bulimia or anorexia nervosa)

ANS: bulimia DIF: I OBJ: 7-1.4

5. A person who uses drugs for their intended purpose and with the proper dosage is a ____________________. (drug abuser or drug user)

ANS: drug user DIF: I OBJ: 7-2.1

6. Chloride and magnesium are examples of ____________________. (minerals or nutrients)

ANS: minerals DIF: I OBJ: 7-1.1

7. The body uses building blocks, called amino acids, to make ____________________. (proteins or carbohydrates)

ANS: proteins DIF: I OBJ: 7-1.1

8. Over-the-counter and prescription drugs may produce unwanted ____________________. (withdrawal symptoms or side effects)

ANS: side effects DIF: I OBJ: 7-2.1

9. ____________________ refers to methods of preserving and protecting your health. (Malnutrition or Hygiene)

ANS: Hygiene DIF: I OBJ: 7-3.1

10. All ____________________ are made from opium. (drugs or narcotics)

ANS: narcotics DIF: I OBJ: 7-2.2

11. ____________________ provide more than twice as much energy as proteins and carbohydrates per unit mass.

ANS: Fats DIF: I OBJ: 7-1.1

12. ____________________ is a fatlike substance found naturally in the body. However, high levels in the blood increase the risk of heart disease.

ANS: Cholesterol DIF: I OBJ: 7-1.1

13. ____________________ are organic compounds that control many body functions, but they cannot be made by the body.

ANS: Vitamins DIF: I OBJ: 7-1.1

14. A ____________________ is any chemical substance that causes a physical or emotional change in a person.

ANS: drug DIF: I OBJ: 7-2.1

15. ____________________ are drugs that speed up the action of the central nervous system.

ANS: Stimulants DIF: I OBJ: 7-2.1

16. ____________________ are drugs that slow down body functions and may reduce a person's alertness.

ANS: Depressants DIF: I OBJ: 7-2.1

17. ____________________ is physical dependence on a drug, and if the body does not receive the drug, withdrawal symptoms may occur.

ANS: Addiction DIF: I OBJ: 7-2.1

18. Sometimes dependence on a drug is not due to a physical need. Some people form a ____ on a drug and feel powerful cravings for it.

ANS: psychological dependence DIF: I OBJ: 7-2.1

19. ____________________ is a chemical stimulant in tobacco that increases heart rate and blood pressure.

ANS: Nicotine DIF: I OBJ: 7-2.2

20. People who suffer from ____________________ are dependent physically and psychologically on alcohol.

ANS: alcoholism DIF: I OBJ: 7-2.2

21. ____________________ is the physical and mental response to pressure.

ANS: Stress DIF: I OBJ: 7-3.3

22. ____________________ is a condition characterized by an extremely high percentage of body fat.

ANS: Obesity DIF: I OBJ: 7-1.4

23. Oranges contain ____________________, which helps your body absorb iron and fight disease.

ANS: vitamin C DIF: I OBJ: 7-1.1

24. Carrots contain ____________________, which helps keep your eyes healthy.

ANS: vitamin A DIF: I OBJ: 7-1.1

25. Your body uses the ____________________ in tortillas and bread for energy.

ANS: carbohydrates DIF: I OBJ: 7-1.1

26. The mineral ____________________, found in milk helps strengthen your bones.

ANS: calcium DIF: I OBJ: 7-1.1

27. ____________________ are nutrients that are NOT made by living things.

ANS: Minerals DIF: I OBJ: 7-1.1

28. A ____________________ shows the kinds and amounts of foods you should eat each day.

ANS: food pyramid DIF: I OBJ: 7-1.2

SHORT ANSWER

1. Name the six groups of nutrients, and explain why each is important to the body.

ANS:
Sample answer: Carbohydrates provide energy to the body. Proteins provide amino acids so the body can make other proteins. Fats are an energy source, and they provide insulation to the body. Vitamins and minerals control many body functions. Water transports substances, regulates body temperature, and provides lubrication.

DIF: I OBJ: 7-1.1

2. If vitamins and minerals do not supply energy, why are they important to a healthy diet?

ANS:
Vitamins and minerals help the body use other nutrients properly, and they control many body functions.

DIF: I OBJ: 7-1.1

3. Name some of the nutrients that can be found in a glass of milk.

ANS:
Milk contains carbohydrates; protein; vitamins A, B_2 (riboflavin), B_{12}, and D; biotin; calcium; magnesium; potassium; and water.

DIF: II OBJ: 7-1.2

4. What information is found on a nutrition information label?

ANS:
The nutrition facts panel includes the serving size, the number of servings per container, the number of Calories per serving, and the percent daily value of each type of nutrient per serving.

DIF: I OBJ: 7-1.3

5. How do anorexia nervosa and bulimia differ?

ANS:
Anorexia nervosa involves self-starvation. Bulimia involves binge eating followed by induced vomiting or the use of laxatives.

DIF: I OBJ: 7-1.4

6. How can someone who is obese suffer from malnutrition?

ANS:
An obese person can suffer from malnutrition by eating too many of the wrong nutrients and too few of the right nutrients.

DIF: II OBJ: 7-1.4

7. What is the difference between a prescription drug and an over-the-counter drug?

ANS:
A prescription drug can only be purchased after your doctor writes a prescription telling the pharmacist what drug and what dosage to give you. An over-the-counter drug can be purchased by anyone without a prescription.

DIF: I OBJ: 7-2.1

8. How does addiction occur?

ANS:
Addiction occurs when a person takes an addictive drug over a period of time and becomes physically dependent on the drug.

DIF: I OBJ: 7-2.1

9. a. How are nicotine, alcohol, heroin, and cocaine similar?
   b. How are they different?

   ANS:
   a. Nicotine, alcohol, heroin, and cocaine are all addictive drugs.
   b. Nicotine and alcohol are legal for use by adults. Cocaine and heroin are illegal. Cocaine and nicotine are stimulants. Alcohol is a sedative. Heroin is a narcotic.

   DIF: II OBJ: 7-2.2

10. What is aerobic exercise? Give three examples.

    ANS:
    Aerobic exercise is vigorous, sustained exercise of the whole body for 20 minutes or more. Some examples of aerobic exercise include biking, in-line skating, jumping rope, and swimming.

    DIF: I OBJ: 7-3.2

11. What should you do when calling for help in a medical emergency?

    ANS:
    Dial 911, speak slowly and clearly, and describe where you are, why you called, how many people were injured, and what types of injuries were sustained. Listen carefully to the instructions, and let the other person hang up first.

    DIF: I OBJ: 7-3.4

12. What is hygiene, and how does it help you stay healthy?

    ANS:
    Hygiene includes methods of preserving and protecting your health. It keeps you healthy by keeping your body in top condition.

    DIF: I OBJ: 7-3.1

13. a. What situations are causing excess stress in your life right now?
    b. What can you do to help relieve the stress you are feeling?

    ANS:
    Answers will vary.

    DIF: II OBJ: 7-3.3

14. What needs must a healthy diet meet?

    ANS:
    It must provide enough essential nutrients and meet the body's energy needs.

    DIF: I OBJ: 7-1.2

15. Why should people strive to limit the fat in their diet?

ANS:
Too much fat in the diet has been linked to weight gain, heart disease, and some kinds of cancer.

DIF: I OBJ: 7-1.2

16. How might a very low-fat diet affect the body?

ANS:
Too little fat might result in hormone imbalances, dry skin, and vitamin deficiencies.

DIF: I OBJ: 7-1.2

17. What can be done to prevent drug abuse?

ANS:
Answers will vary but should mention that information about the risks associated with drug use is an effective weapon. Being well informed is the first step toward making informed choices.

DIF: I OBJ: 7-2.1

18. How are nicotine and alcohol alike? How do they differ?

ANS:
Answers will vary but should reflect the understanding that both drugs are legal for adults and that both are addictive. The use of both nicotine and alcohol has been associated with heart disease. The primary physical consequences of smoking are respiratory and cardiac; alcohol mostly affects the central nervous system, the liver, and the digestive system. Of the two, only alcohol dangerously impairs judgment.

DIF: I OBJ: 7-2.2

19. How are healthy habits related to stress?

ANS:
Practicing healthy habits, such as exercising regularly and getting adequate rest, helps us cope with stress in a positive way.

DIF: I OBJ: 7-3.3

20. What do bicycle helmets and seat belts have in common?

ANS:
They both decrease the risk of injury from accidents.

DIF: I OBJ: 7-3.4

21. When you call for emergency help, what should you tell the person who answers the phone?

ANS:
Tell the person the location of the emergency, the nature of the emergency, the number of people injured, and the type of injuries involved. Always hang up last.

DIF: I OBJ: 7-3.4

22. Are all narcotics illegal? Explain.

ANS:
Not all narcotics are illegal. Some narcotics are prescription drugs used to treat severe pain.

DIF: I OBJ: 7-2.2

23. a. What are the three types of nutrients that provide energy in Calories?
    b. Which three nutrients do not provide energy?

ANS:

a. The three nutrients that provide energy in Calories are carbohydrates, proteins, and fats.
b. Three nutrients that do not provide energy are water, vitamins, and minerals.

DIF: I OBJ: 7-1.1

24. If you drink a beer, are you taking a drug? Explain your answer.

ANS:
Beer contains alcohol, which depresses the central nervous system. Alcohol is a drug. Therefore, if you drink beer, you are taking a drug.

DIF: I OBJ: 7-2.2

25. Use the following terms to create a concept map: *unsaturated fats, carbohydrates, water, proteins, nutrients, simple sugars, starch, fats, vitamins, minerals, saturated fats.*

ANS:

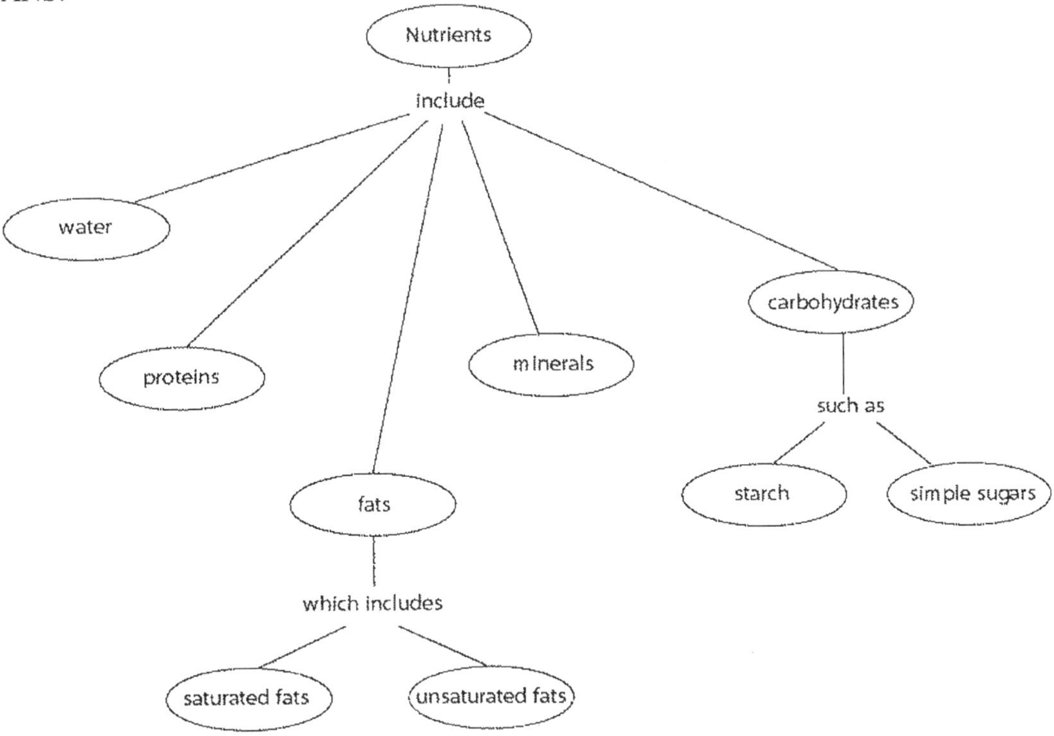

DIF: II OBJ: 7-1.1

26. Many people eat little or no meat. Meat is often the major source of protein in American diets. What other sources of protein can vegetarians choose?

ANS:
Sample answer: Other foods that contain protein include eggs, dairy products, and fish.

DIF: II OBJ: 7-1.2

27. You are at a party and a friend offers you a cigarette. He says that one cigarette won't hurt you. Using what you know about tobacco and addiction, explain why his reasoning is false.

ANS:
Sample answer: The nicotine in cigarettes is highly addictive. One cigarette could easily lead to many more.

DIF: II OBJ: 7-2.2

28. Assume your diet is 20 percent fat, 30 percent protein, and 50 percent carbohydrates. In order to get 2,000 Calories a day, how many Calories of each nutrient should you be getting?

ANS:
You should get 600 Cal from protein, 400 Cal from fat, and 1,000 Cal from carbohydrates.

DIF: II OBJ: 7-1.2

29. What are essential vitamins?

ANS:
Essential vitamins are organic compounds that control many body functions. Most essential vitamins must be obtained from sunlight and by eating the right foods.

DIF: I OBJ: 7-1.1

30. List food sources for five essential vitamins.

ANS:
Sample answer: Vitamin A is in milk. Vitamin C is in citrus fruits. Vitamin D is in eggs. Vitamin E is in liver. Vitamin K is in leafy greens.

DIF: I OBJ: 7-1.1

31. Why do you think only packaged foods are required to have nutritional information labels?

ANS:
Sample answer: Packaged foods are required to have nutritional information labels because they have been processed. Packaged foods may lose nutrients during processing, and they may contain additives or preservatives. The nutritional content of foods that are not packaged, such as produce, can be easily determined because they have not lost or gained ingredients.

DIF: II OBJ: 7-1.3

32. Use the following terms to complete the concept map below: *cocaine, heroin, marijuana, narcotic, opium, stimulant, Indian hemp plant.*

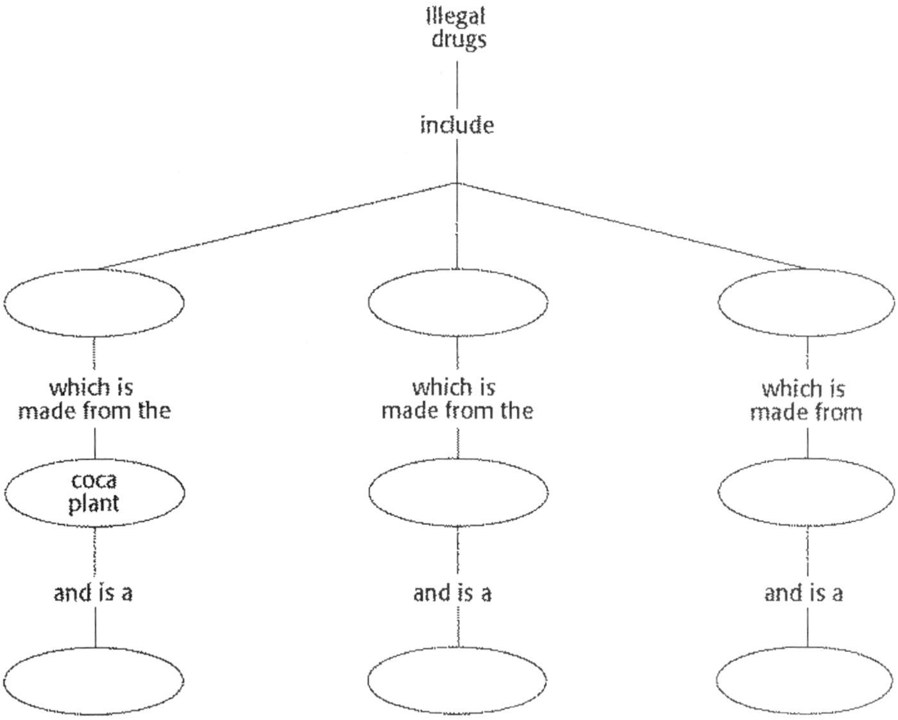

ANS:

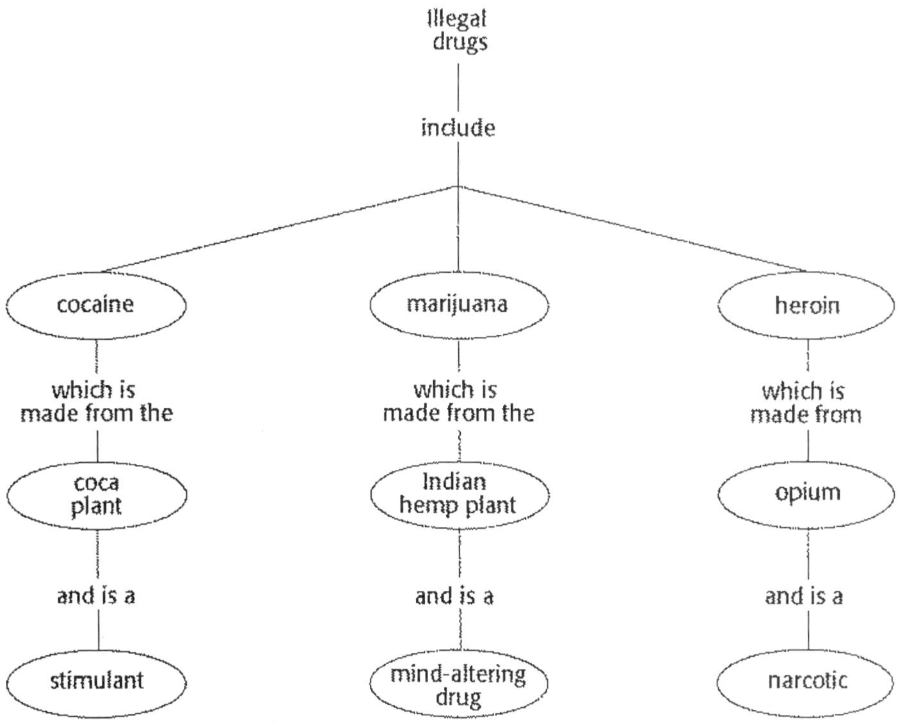

DIF: II OBJ: 7-2.2

Examine the picture below, and answer the items that follow.

33. a. Name four unhealthy practices in the picture above.
    b. Name four healthy practices in the picture above.

ANS:

a. Sample answer: The people in the boat are not wearing life jackets. The swimmer is swimming alone. The bike rider does not have a helmet on. People are in the water while a thunderstorm is approaching.
b. Sample answer: The hikers are exercising together. The hikers have water with them. The biker is putting on sunscreen. The hikers, biker, and swimmer are all performing aerobic exercise.

DIF: II OBJ: 7-3.4

34. How are tolerance, addiction, and withdrawal symptoms related?

ANS:
Regular use of some drugs may cause the body to develop tolerance. Larger and larger doses of the drug are needed to get the same effect. With some drugs, even a single use could lead to addiction, in which the body is dependent on the drug. Withdrawal symptoms occur when the body does not receive a drug that it is addicted to.

DIF: II OBJ: 7-2.1

35. Why might planning for a school project help to reduce stress?

ANS:
Planning helps you realize exactly what tasks need to be done and how long you have to do them. You might feel better knowing that there was a certain time in which to do each task so that the entire project will be finished on time.

DIF: II OBJ: 7-3.3

36. Your friend drinks two cans of caffeinated soft drink every day. One day your friend is out of his soft drink and does not drink his usual two cans. He begins to feel tired and nauseous, and he develops a headache.

ANS:
Sample answer: My friend's body has become accustomed to (and possibly addicted to) the caffeine in the soft drink. When he did not drink his usual amount of soft drink, he experienced withdrawal symptoms due to the lack of caffeine.

DIF: II OBJ: 7-2.1

37. Why would a diet without any protein be dangerous to your health?

ANS:
The body would have no way to grow, repair itself, or produce new proteins.

DIF: II OBJ: 7-1.1

38. Each week for a month, an overweight person used more Calories exercising than were obtained from the food he or she had eaten that week. Compare the amount of fat tissue in the individual's body at the end of the month with the amount of fat tissue at the beginning of the month. Explain any difference.

ANS:
At the end of the month, the individual should have less fat tissue than he or she had at the beginning of the month. Fat tissue was lost as the stored energy was used during exercise.

DIF: II OBJ: 7-1.1